HEINKEL He 219

HEINKEL
He 219

An Illustrated History of Germany's Premier Nightfighter

Roland Remp

Schiffer Military History
Atglen, PA

Book Design by Ian Robertson.
Translated from the German by David Johnston.

Copyright © 2000 by Schiffer Publishing, Ltd.
Library of Congress Catalog Number: 00-105622

Printed in China.
ISBN: 0-7643-1229-4

This book was originally published under the title,
Der Nachtjäger Heinkel He 219
by Aviatic Verlag GmbH

We are interested in hearing from authors with book ideas on related topics.

Published by Schiffer Publishing Ltd.
4880 Lower Valley Road
Atglen, PA 19310
Phone: (610) 593-1777
FAX: (610) 593-2002
E-mail: Schifferbk@aol.com.
Visit our web site at: www.schifferbooks.com
Please write for a free catalog.
This book may be purchased from the publisher.
Please include $3.95 postage.
Try your bookstore first.

In Europe, Schiffer books are distributed by:
Bushwood Books
6 Marksbury Avenue
Kew Gardens
Surrey TW9 4JF
England
Phone: 44 (0) 20 8392-8585
FAX: 44 (0) 20 8392-9876
E-mail: Bushwd@aol.com.
Free postage in the UK. Europe: air mail at cost.
Try your bookstore first.

Contents

Introduction **6**

Acknowledgment **7**

The History of the Heinkel He 219 "UHU" **8**

Walkaround the He 219 Prototype **29**

Testing **46**

Appendices **133**

Introduction

Two things prompted me to write this type history: one was the reconstruction of the crash of an He 219 near my birthplace in which I was involved from an early age, the other the fact that published information concerning this interesting aircraft was sketchy at best.

I selected a chronological account as the best way of portraying the history of the He 219. Even though this deviates from the standard set in other such publications, it seemed to me better suited for describing the technical development and the unbelievable delays in the machine's service introduction, which resulted from the tactics of the RLM. For this type was the subject of constant modification and redesign, especially during its brief service life, more so than almost any other type. In spite of everything, however, the He 219 proved over and over that it had little to fear from any German-designed competitor—at least in 1944-45. The development of turbojet-powered night fighters was still in its infancy, and existing piston-engined machines could not look forward to improved engines as there were none. Consequently, the advantages—lightness of construction, cost of manufacture, ease of maintenance and handling, and not least outstanding flying characteristics—remained on the side of the He 219.

Many sources, especially from the period after the autumn of 1944, were lost or destroyed. I am certain that some remain hidden in archives or private collections all over the world. As a result, I cannot claim that this account is error free or definitive. I would, therefore, be very grateful to the reader for any corrections or additional information which could be included in subsequent editions.

Acknowledgment

Since photographs of the He 219 are relatively scarce, providing new, unpublished photos for this book proved extremely difficult. Some of the photos which appear in this book will surely be familiar to a number of readers, however, for the sake of completeness their use was unavoidable. Other photographs were generously made available to me by members of I./NJG 1 or comrades of the *G.d.J. Vereinigung der Flieger Deutscher Streitkräfte e.v.*

I would like to express my appreciation to the former crews of I/NJG 1, who provided accounts and invaluable support in the preparation of this book. In particular I would like to mention Heinrich Becker, Prof. Dr. Ing. O.H. Fries, Fritz Habicht, Hans Höhler, Ing. Georg Kösel, Dr.Ing. Herbert Kümmritz, Herbert Scheuerlein (deceased) and Dr. Helmut Zach. Valuable assistance was also provided by Mr. Peter Murton, who made available highly interesting documents from the archives of the Royal Air Force Museum, Mr. Tom Alison of the National Air and Space Museum in Washington, D.C., Hans Scholl, Harold Thiele, Walter Shick (deceased) of the Heinkel Company, and Günter Sengfelder. I would also like to mention my dear wife and my doctor, as well as my friend and mentor, Ingolf Meyer, who encouraged me to complete this book even though it meant that all too often I was apart from them.

Roland Kemp
March 1999
89335 Ichenhausen
Lapierstrasse 4
E-Mail Address: Rremp71690

The History of the Heinkel He 219 "UHU"

Project Phase

The steadily growing threat posed by the Royal Air Force's Bomber Command forced the RLM to create an independent night fighter force. Consequently, the order for the formation of a night fighter division (*Nachtjagddivision*) was issued on 17 July 1940. Its commander was *General* Kammhuber, a man who had amassed considerable night flying experience with KG 51. Dissatisfied with the aircraft types then available (versions of the Bf 110 heavy fighter and the Do 17 high-speed bomber converted for the night fighter role), soon after assuming his new post Kammhuber called for a specialized night fighter. The twin-engined machine proposed by him would possess good handling characteristics, a crew of two seated side-by-side, extensive glazing to provide good visibility for the crew, and fixed cannon armament in a fairing beneath the fuselage (to shield the crew from the muzzle flash).

A few months earlier, on 28 April 1940, the Heinkel factory in Rostock-Marienehe had on its own initiative submitted a proposal to the RLM for a high-speed single-engined reconnaissance aircraft. Later, this proposal would form the basis of the most outstanding night fighter of the Second World War. At that time, however, no one gave any thought to this future operational role. Reconnaissance machines and bombers were high on the priority list, and a firm like Heinkel naturally strove to acquire contracts for types which would be built in large numbers. Project work was based on a declaration by the RLM that the *Luftwaffe* currently possessed no efficient reconnaissance machines, which meant that the reconnaissance role had to be filled by aircraft for the reconnaissance mission as a secondary role. The General Staff, therefore, demanded an aircraft designed exclusively for the reconnaissance mission whose performance would not be hampered by the need to fill other roles, such as carrying bombs.

Initial discussions between representatives of the RLM and director Lusser, Herr Ebert, and Herr Meschkat of the Heinkel Company were held at the RLM on 30 September and 1 October 1940. The Heinkel representatives outlined their newest project, a reconnaissance aircraft based on the He 119 with the designation P 1055.

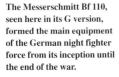

The Messerschmitt Bf 110, seen here in its G version, formed the main equipment of the German night fighter force from its inception until the end of the war.

8

The Ju 88 C, here equipped with FuG 202 BC radar, formed the second leg of the German night fighter defense until 1943.

The proposed machine had a wing area of 42 square meters, a range of 4,000 kilometers, a maximum speed of 750 km/h, an all-up weight of 12.6 tons, and a takeoff distance of 840 meters.

Herr Lusser proposed interchangeable outer wings, which would make it possible to select a wing area of 35 to 45 square meters. In an effort to increase the RLM's interest, the Heinkel representatives suggested a possible armament of two 1,000 kg bombs carried on external racks. The aircraft was to have a crew of two in a pressurized cockpit. Herr Friebel (RLM LC 2 C) held out the prospect of a binding contract in the very short term, with maximum speed the determining factor. Inquiries were made about performance figures for a version powered by the DB 613 with gas-

driven turbo supercharger. This engine was under test by Daimler Benz and was supposed to become available in 1941. The new power plant, which consisted of two DB 603 engines driving a single shaft with turbo supercharger and methanol-water injection, was supposed to deliver 3,500 hp for takeoff. The prospect of this engine put thoughts of true high-altitude flight into the heads of planners in the RLM and the aviation industry.

By the time the next conference with the RLM took place on 24 October, there had been several discussions and an exchange of letters between senior Heinkel staff members, who were far from being of one mind about the new aircraft project. These letters reveal that Heinkel was striving towards a new major contract. Interest-

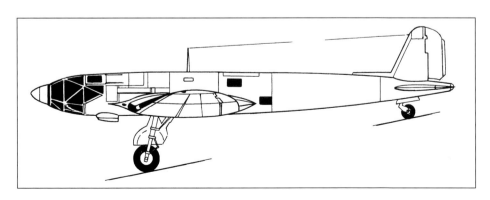

Project P 1055 was based on the Heinkel He 119, which was powered by coupled engines placed in the center of the fuselage.

ingly, considerable doubts were expressed about the advisability of a new aircraft which, based on recent experience, would probably not reach the units until mid-1944, by which time it would have been rendered obsolescent by newer designs, possibly equipped with turbojet engines. Potential use of the type as a bomber, which had proved appealing to the RLM at the last conference, produced serious concerns which were based on a number of considerations. Forward defense was considered inadequate, since the propeller restricted forward-firing armament to a fixed installation. Heinkel was aware of the heavy losses sustained by He 111 units over England, and the reason for this was seen quite clearly as the type's woefully inadequate armament and its extensive nose glazing. Since the P 1055 would also fly in formation, a fixed forward-firing armament would be fatal. The crew of two was also seen as inadequate, since one gunner could operate just one weapon, and if he were knocked out the aircraft would become easy prey for an attacker. While Director Lusser was of the opinion that the P 1055 would be in little danger of attack on account of its high speed, Director von Pfistermeister tried to describe his vision of the future:

"According to its design planning, the P 1055 will take to the air at the end of 1942. If the first machines then reach the units in October 1944, four years will have passed since the present day. I cannot agree with your view that we can reckon with some certainty that the P 1055 will retain its great advance in speed until that time. By then there might be fighters with turbojet engines which will far surpass the 750 km/h speed of the P 1055. I even hope that these will be based on the Heinkel 280 fighter. Then the P 1055's need for greater defensive capabilities will be exactly the same as that of present day bombers. I consider it a mistake to ignore the possibility, if not the likelihood, of increased speeds for fighters."

Following on the heels of these and other objections, discussions were held on the use of the P 1055 as a medium bomber, which had been designated the most important type at that time. This would, however, require design modifications in the areas of size, load-carrying ability, crew, and defensive positions. It was feared, however, that the RLM would surely give preference to a bomber with stronger defensive capabilities, even if this came at the cost of reduced speed. Special mention was made of the contemporary Ju 288, which was heavily armed and had a maximum speed of 650 km/h, in comparison to the P 1055.

Director von Pfistermeister invited Herr Friebel of the RLM to dinner and used the opportunity to discuss outstanding questions. The General Staff considered the broadening of the range of roles of reconnaissance aircraft to be vitally important. Reference was only made to postwar tasks in Africa, to the photo-mapping of unexplored regions, and to the Near and Far East. At this meeting Friebel advised that Heinkel should stick stubbornly to the task of producing a first-class reconnaissance machine. He considered a bomber based on the P 1055 to be completely unnecessary and unsuitable, since the He 177 was to become one of the Luftwaffe's most important types, while the idea of producing a Heavy fighter (Zerstörer) based on the same design was seen as questionable. The catastrophic losses suffered over England by the Bf 110 and Me 210 were cited, as these were inferior to regional fighters even without a bomb load.

In spite of all objections, specifications were calculated for presentation to the RLM (see Table 1) for the heavy fighter and bomber roles.

Herr Reidenbach of the RLM, with whom the discussion of 19 October 1940 had been held, noted that the RLM would

Wing area	37 m²
Range	3 000 km for destroyer and bomber
Maximum speed	745 km/h at 2,000 h.p. at 6 000 m in destroyer role
Gross weight	11 000 kg in destroyer role
Bomb load in dive-bomber role	2 x 1 800 kg
Fixed armament in destroyer role	4 MG 151 (2 in fuselage, 2 in wings)

Table 1

Wing area	38.5 m²
Range	2 000 km (full throttle)
4 000 km (cruise power)	
Maximum speed	735 km/h, 2,000 h.p. at 6 000 m
Weight	11 100 kg
Takeoff distance	780 m
Service ceiling	9 800 m at 1,700 h.p.

Table 2

Wing area	37 m2
Maximum speed	720 km/h, 3,200 h.p. at 9 000 m
Range	2 000 km
Armament	2 MG 151 fixed forward-firing
	1 MG 131 Z in dorsal position
	1 MG 81 Z in ventral position

Table 3

initially only be able to acknowledge the information. A decision on this would be left open for the time being. Herr Christensen, who was also present, also pointed out that, after the experience with the Bf 110, the heavy fighter would really only make sense if it was in the same speed class as single-seat fighters. For further work on the project, Heinkel received a directive that the reconnaissance version should receive the project designation P 1055 and the heavy fighter P 1056.

At the main conference on 24 October 1940 the Heinkel Company presented a fuselage cross-section of the projected machines. The heavy fighter version showed the use of a nosewheel undercarriage for the first time. Associated technical data were given (see Table 2).

On the whole the RLM was in agreement with the project, however, the following improvements were considered absolutely necessary:

a) Service ceiling raised to 12,500 m.

b) Further increases in service ceiling through the use of exhaust-driven turbo superchargers.

c) The armament package proposed by Heinkel—one MG 151 in a remotely-controlled dorsal turret and two fixed forward-firing MG 151s—was acknowledged as adequate. However, defensive armament was to keep pace with increased speeds of enemy fighters, which resulted in a demand for remotely-controlled turrets above and below the rear fuselage.

d) Variable wing area achieved through the use of interchangeable outer wing panels.

e) Maximum dive angle of 30°.

f) Wing de-icing equipment, which would be especially vital for operations over the Atlantic.

g) Propeller blades to be sized to accommodate greater engine outputs.

Wing area	45 m²	
Maximum speed	700 km/h, 3,200 h.p. at 9 000 m	
650 km/h, 3,700 h.p. at 6 000 m		
Gross weight	11 400 kg (medium weight)	
Service ceiling	10 500 m	
Range	3 500 km	

Table 4

Wing area	42.5 m²	50 m²	57.5 m²
Maximum speed at 9 000 m	686 km/h	668 km/h	653 km/h
Service ceiling (max. weight)	9 800 m	10 250 m	10 700 m
Landing speed	168 km/h	156 km/h	148 km/h

Table 5

Data were also provided for the proposed Zerstörer project P 1056 (see Table 3).

For subsequent development the RLM proposed the use of a total of six remotely-controlled turrets, three above and three beneath the fuselage. Furthermore, the RLM asked that development of the *Zerstörer* be continued with a bolstered armament of at least three rotating gun turrets above and beneath the fuselage. The objective was to create a force of heavily armed fighters which would form a "hedgehog" around the bombers.

Inspection of the P 1055 mockup took place on 23 November 1940, and the RLM representative, Herr Scheibe, approved in principle the arrangement of the crew, visibility and space conditions, armor, and the arrangement of periscopes. Interestingly, the RLM now desired that the project be developed as a reconnaissance machine, as a day bomber, and as a Heavy fighter. The bomber version was to carry its payload in a fuselage bay, rather than on external racks. Director Lusser went further and proposed that the P 1055 be equipped with two twin-gun barbettes above the fuselage and two below, giving it a total of eight movable weapons plus one or two fixed, forward-firing guns.

On 28 November 1940 the RLM announced that the P 1055 reconnaissance bomber had been added as an element of the ministry's development planning. Deadline for the final mockup inspection was set for 15 January 1941, however, not much hope was held out for the issuing of a contract for the P 1056 Zerstörer.

At this point the Ernst Heinkel Flugzeugwerke introduced a number of innovations: adoption of a mid-wing monoplane arrangement, relocation of the engine to a position in front of the wing for better accessibility, a more favorable radiator arrangement, improved forced air induction, and the ability to carry either long-range tanks or bombs for the strategic reconnaissance and bomber roles, respectively. (For technical specification see Table 4.)

Increasing wing area by 20 percent was expected to result in an increase in service ceiling of about 1,200 meters. For the day bomber role, Heinkel declared that the aircraft would be able to carry a payload of one 1,000-kg bomb, two 500-kg bombs, or two 250-kg bombs.

On 13 December 1940 another conference was held, attended by representatives of the RLM, the *Erprobungsstelle Rechlin*, and the Ernst Heinkel Flugzeugwerke (EHF). Heinkel produced further performance calculations for the P 1055 with DB 613 motors (see Table 5).

The problem of the twin tail arrangement was discussed again: it was feared that variations in the air flow behind the propeller might produce vibrations in the tail surfaces. The proposed solution was the adoption of a single fin and rudder. The "armed hedgehog" concept was also discussed again. For aerodynamic reasons, the RLM requested that the barbettes be retractable. The anticipated weapons arrangement was one turret housing three MG 151s and one housing four MG 131s above the fuselage, and a single turret housing four MG 131s beneath the fuselage.

By now work on the project had reached the stage where a type designation for the future machine had to be found. The Dornier Company had already claimed the designation Do 219 for a development of the Do 17, but nevertheless the RLM assigned the same number to Heinkel, and the project received the designation He 219. From this point on the P 1055 became the He 219.

Heinkel was advised that the representatives of the RLM and the *Erprobungss-*

telle Rechlin would visit the Ernst Heinkel Flugzeugwerke to inspect the mockup of the He 219 on 4 February 1941.

A Heinkel memorandum dated 13 February 1941 compared the performance of the He 219 at various stages in its development with that of the Focke Wulf Fw 191 and the Junkers Ju 288. All performance figures were estimated by Heinkel (see Table 6).

Further discussions between representatives of the RLM and Heinkel were held in mid-February 1941. Herr Friebel of the RLM expressed the view that, based on the data in Performance Sheet No. 1277, the He 219 was at the lowest, though perhaps still acceptable, performance level. The RLM demanded a maximum speed of 750 km/h, which is what Heinkel had promised in its first project submission. It already held the view that a reconnaissance machine must have reached its optimum performance when it arrived at the front, rather than being designed for further increases in performance, as was the case with the bombers. Speed was the most important factor for use in the reconnaissance role. Armament was to be reduced as required to keep pace with advances in the speeds of

Performance	He 219				Fw 191	Ju 288
	30/09/40	19/10/40	28/11/40 D.-Blatt 1274	24/1/41 D.-Blatt 1277		
Max. speed (km/h)	750	735	700 at 9 000 m 650 at 6 000 m	680 at 9 000 m 630 at 6 000 m	590	620
Weight (kg)	12 200	12 100	12 500 recon. 13 500 bomber	14 200 recon. 14 200 bomber	22 500	18 500
Range without bombs optimal (km)	4 000	4 000	2 700	3 300	5 000	5 000
Service ceiling without bombs (m)	?	9 800	10 500	10 700	?	?
Bomb load (kg)	2 000	1 000	1 000	1 000	2 000	2 000
Wing area (m2)	35-45	?	45	55	70	60

Table 6

enemy fighters. The existing design did have one definite advantage, especially if turbojet fighters appeared: against such fighters only armed reconnaissance by an aircraft with a powerful armament and heavy armor made any sense.

Meanwhile, Kammhuber's *Nachtjagddivision* had been expanded into the *XII. Fliegerkorps*, and he was made *General der Nachtjagd* (General in Command of Night Fighters) with the rank of *Generalleutnant*. By the spring of 1941 Bomber Command was attempting to weaken the German armaments industry by launching heavier night raids. In spite of this, the RLM had not yet begun thinking about the requirement for a modern night fighter. Its attention was directed elsewhere, including toward a new reconnaissance aircraft. In an attempt to achieve the maximum possible performance, the RLM made the following proposals for Heinkel's projected reconnaissance machine:

- **Elimination of a possible internal bomb bay.**
- **Reduction of wheel size to 1,100 mm diameter.**
- **Elimination of Fowler flaps as a weight saving measure.**
- **Two-man cockpit.**
- **Maximum speed of 750 km/h.**
- **Elimination of the nosewheel undercarriage if this results in space savings and improved performance.**

As well, Heinkel received the correct performance figures for the Ju 288 and Fw 191 so as to be able to correct its planning (see Table 7). Just two weeks later, Heinkel submitted its recalculated figures to the RLM (see Table 8).

The old project was subsequently dropped, and instructions were issued for a visual mockup and a design mockup of the new project. The lowest acceptable maximum speed was 730 km/h. The RLM requested that the first prototype be ready to fly by 1 March 1942 and the first production aircraft by June 1942.

The He 219 was to be powered by two DB 613 power plants, each driving a three-blade propeller. The two propellers were to turn in opposite directions to reduce torque effect. There was an express requirement that the radiators be protected, and, if possible, fed with forced air. The use of annular radiators with forced air delivered by a fan attached to the propeller shaft (a la Fw 190 with BMW 801) appeared to be the most favorable arrangement.

	Ju 288	Fw 191
Wing area	60 m^2	70 m^2
Takeoff weight, recon. role	17 700 kg	20 500 kg
Fuel	5 500 kg	6 030 kg
Bombs	2 000 kg	2 000 kg
Maximum speed at 6 000 m at combat power (Jumo 222) 2 x 1,700 h.p.	595 km/h	565 km/h

Table 7

Crew	2
Wing area	42.5 m^2
Weight	12 000 kg
Armament	1 x MG 131, fixed rearward-firing
dorsal and ventral turrets each with one MG 131 Zwilling	
Maximum speed	740 km/h (at 9 000 m with emergency power)
Range	2 800 km (optimal)

Table 8

Special crews were to be equipped to fly this special aircraft. The He 219 was to receive priority over current production aircraft for the best equipment. The aircraft mockup was inspected on 26 March 1941. The cockpit arrangement was judged to be very satisfactory. A second sight for the observer's periscope was demanded, so as to enable the pilot to monitor the entire area behind the aircraft. Division of responsibilities for the crew was as follows:

Pilot: In addition to flying the aircraft, the pilot is to scan all the airspace behind the aircraft and watch for enemy aircraft. He operates the fixed weapons with the help of a crosshairs in the fixed periscope.

Observer: Totally responsible for picture taking, he constantly monitors the terrain and corrects the course with the planned autopilot override. When threatened by enemy aircraft, he operates the two gun turrets (*B-* and *C-Stand*).

Provision of a periscope was linked to the requirement that the pilot be able to observe all the airspace around the aircraft: a rearwards-facing periscope with swiveling head for the observer (with view port) for observation of the airspace behind, above, and below, and for aiming the defensive weapons (*B-* and *C-Stand*). The periscope, which projected from the fuselage, was to be enclosed in a teardrop-shaped fairing.

Interestingly, it was anticipated that the He 219 would see continued use as a high-altitude reconnaissance aircraft after the war. The following is from the report on the mockup inspection and subsequent discussion: "This role is dependent on making the aircraft impossible to detect at high altitude. An increase in ceiling can easily be achieved by:

1. eliminating the gun positions, resulting in a weight reduction of approximately 1,000 kg.

2. replacing the existing wing with a very high aspect ratio wing.

Delivery of engines was a determining factor in the delivery date of the first He 219 aircraft. Of the eight DB 613 engines which were supposed to be delivered by the end of 1941, only three were earmarked for Heinkel. The remaining five units were to be tested by Daimler Benz and at Rechlin. A further delay of two to three months was announced by Daimler Benz, because of necessary shaft modifications for the He 219 and He 177.

In addition, the RLM demanded that the He 219 be capable of landing on water. Night operations were to be possible, however, this was not made a condition. A barrage balloon cable fender was foreseen as an equipment set for post-delivery installation. De-icing equipment was definitely supposed to form part of the type's standard equipment.

Periscope trials with a Bf 110 produced such poor results that the He 219 program was placed in danger of cancellation. For this reason a new three-seat design was developed; the sole purpose of the third crew member was to visually monitor the airspace around the aircraft.

On 30 May 1940 there was another conference in Berlin. Since it appeared unlikely that the DB 603 would be delivered by the end of 1942 in time for the first twenty He 219s, an interim solution was proposed with DB 610 engines, a smaller wing, and reduced armament. A further project was proposed with a DB 610 installed in the forward fuselage. A simple mockup was ordered for this new project. Interestingly, the RLM also proposed that the Arado Ar 240 be evaluated in the long-range reconnaissance role. Three prototypes had already been built, and the type was seen as a possible interim solution until the He 219 was ready for production. At the conference Heinkel put forward the latest technical specifications for the He 219 (see Table 9).

Jettisonable long-range tanks were anticipated, the tanks used by the Me 210 having proved especially advantageous.

The RLM proposed replacing the paired MG 131 with two MG 151s for rearwards defense on account of the latter's better range. Herr Friebel mentioned in passing that a special high-altitude reconnaissance aircraft (for altitudes to about 15,000 m) was in accelerated development.

Mockup inspection took place in the Heinkel factory at Marienehe on 20 June 1941. Heinkel proposed two solutions:

Solution 1: a version of the He 219 with DB 613 engines

Solution 2: a new aircraft with one DB 610 coupled engine in the forward fuselage.

Since Solution 1 resulted in poorer visibility for the pilot and Solution 2 of-

fered significant advantages for observation and defense to the rear and below, it was decided that Heinkel would produce a mockup of Solution 2 with a DB 615 engine by 12 July 1941. Use of this power plant was expected to result in a maximum speed of 760 km/h at an altitude of 10,000 meters. A pressurized cockpit was demanded for Solution 2. An ERKU CI telescope produced by the Zeiss Company was to be tested in one of the weapons stations of an He 111.

The RLM abandoned the interim solution of a DB 610 powered He 219 and replaced it with the Ar 240. On 25 June 1941 the RLM finally dropped the idea of a version with a fuselage-mounted engine.

On 11 July 1941 *Generaloberst* Ernst Udet met with Prof. Heinkel, and among other things asked about the He 219 project

He 219 with DB 613 and C3 fuel	
Wing area	42 m²
Armament	two MG 131 Z, remotely-operated
Gross weight	12 200 kg
Maximum speed	693 km/h (emergency power)
at altitude of 6 800 m	667 km/h (combat power)
Range	2 940 km at 490 km/h

He 219 with DB 613, smaller wing	
Wing area	37 m²
Armament	two MG 131 Z, fixed rearward-firing
Gross weight	10 600 kg
Maximum speed	738 km/h (emergency power)
at altitude of 6 800 m	712 km/h (combat power)
Range	2 940 km at 528 km/h

Similar design with DB 6109	
Wing area	32 m²
Armament	two MG 131 Z, fixed rearwards-firing
Gross weight	9 100 kg
Maximum speed	719 km/h (emergency power)
at altitude of 6 800 m	704 km/h (combat power)
Range	2 720 km at 516 km/h

He 219 with DB 610 in forward installation	
Wing area	32 m²
Armament	two MG 131 Z, fixed rearwards-firing
Gross weight	9 100 kg
Maximum speed	740 km/h (emergency power)
at altitude of 6 800 m	724 km/h (combat power)

Table 9

and the search for an engine arrangement. Heinkel had to tell him that design experiments with engines mounted in the rear fuselage or in an extended forward fuselage with revised cockpit had failed to produce a better or satisfactory solution. At this time the entire He 219 project was in serious jeopardy. Udet intended to visit the Heinkel factory at Marienehe on 17 July and take the opportunity to inspect the He 219 mockup. Udet knew of the growing concerns of the Defense of the Reich's night fighter arm and *General* Kammhuber's desire for a specialized night fighter. During inspection of the mockup, talk must have turned to the aircraft's use as a night fighter, for the conference minutes of 18 July 1941 show activities in this direction by the responsible parties for the first time.

Heinkel obtained information from the RLM about night fighting procedures and current night fighter aircraft. This obviously brought new and promising perspectives for the He 219. The RLM stated that the Bf 110 and Do 215 had proved quite capable of dealing with the British Whitleys, Hampdens, and Wellingtons, however, since Bomber Command had introduced the Manchester and the four-engined Stirling and Halifax, successful interceptions had become more difficult. Alarmed by the declining success of his night fighter arm, at about the same time *Generalmajor* Kammhuber requested a meeting with Hitler to convince him of the urgent need to develop a more potent night fighter. Impressed by Kammhuber's arguments and his tenacity, Hitler gave his general broad special powers which made him largely independent of Groehler, the GL/C, and the RLM. Kammhuber acted immediately, and in the period that followed sent successful night fighter pilots like Streib, Lent, and Becker to Rostock to check out the promising Heinkel project, about which Udet had probably informed him, and support Heinkel.

The first reference to the He 219 as a night fighter was made in a memorandum dated 31 July 1941. In this form the aircraft was to be powered by DB 613 or DB 615 engines. Consideration was still being given to the roles of heavy fighter and fast daylight bomber. Out of countless requests from the units arose the requirement for the heaviest possible armament, six cannon at least. This would produce great difficulties with the planned six-blade propellers. Safely firing such a concentration of weapons would be almost impossible. At that time ammunition ignition precision varied from 3.5 to 6 milliseconds. The six-blade propeller offered a free angle of about 30 degrees, with a projectile passage interval of 5 to 7 milliseconds. Thus, the dispersion of the projectiles alone took up this valuable time. Testing of a four-blade propeller also failed to produce satisfactory results. A third crew member was considered desirable for night fighter operations, with at least two crew members sitting side-by-side. Finally, in an effort to save the project, the only correct step was considered, namely fitting the machine with wing-mounted engines. The only negative aspect of this plan was the continued delays with the DB 613 and DB 615 engines. Temporarily equipping the machine with the less powerful but available DB 603 appeared to promise success.

As a result of the problems raised by the fixed forward-firing armament and dive-bombing with the fuselage-mounted engine, on 14 August 1941 Heinkel requested *Generaloberst* Udet's approval to produce a version of the He 219 with two wing-mounted engines. The originally calculated maximum speed would only be reduced to 709 km/h, since the fuselage could now be made significantly slimmer. The use of a high-aspect-ratio wing would permit the machine to reach altitudes of 16,000 meters, though at the cost of a further loss in speed. Udet gave his approval

The DB 603 engine was the standard power plant in almost all variants of the He 219.

The DB 603 engine was the standard power plant in almost all variants of the He 219.

and limited Heinkel to high-altitude reconnaissance, high-altitude fighter, high-altitude night fighter, and night fighter versions of the He 219, probably with the Ar 240 and Me 210 in mind. Heinkel felt that it had an advantage over these competitors, because the now planned armament and power plants meant that Arado and Messerschmitt would also have to redesign their projects.

Heinkel noticed that the countless He 219 project proposals worked out by his company and submitted to the RLM within the year were no longer receiving necessary attention. Even though they felt themselves in a position to also cover the heavy fighter and fast bomber roles, it seemed advisable to discuss these roles again with the RLM when the time came. Thus, these versions of the He 219 project were put on ice indefinitely.

Although the designation He 219 had been assigned, Heinkel attempted to obtain from the RLM a different, possibly higher number. This was probably because the company no longer considered the new twin-engined machine a development of the He 219. Choices were the RLM numbers 274, 276, 277 (reserved for developments of the He 177), 278, and 279. On 22

August 1941 Dr. Heinkel advised the RLM in a telex: "If the designation He 250 is not possible, any other number is alright with me." No change was forthcoming, and the designation remained the He 219.

Heinkel presented the new twin-engined design to the RLM on 26 August. At the request of the ministry the flexible armament was moved farther forward. Also changed was the main undercarriage, each element of which now had one large wheel in place of two smaller wheels. The revised undercarriage retracted flat into the wing without affecting torsional rigidity. As a result of a request by Kammhuber transmitted via the RLM, a three-man cockpit was worked out. Mockup inspection was scheduled for the beginning of September.

In the records of the Ernst Heinkel Flugzeugwerke are secret minutes dated 13 October 1941 concerning a conference with the General Staff held at Marienehe on 10 and 11 October 1941. They state: "...the officers visited Ernst Heinkel Flugzeugwerke to ascertain the status of the He 111, He 177, He 219, and He 280. The opinions of the officers (Maj. Storp, Hptm. Hilscher, Hptm. von Ditfurth, Oblt. Fischer, Fl.St.Ing. Utech, Fl.St.Ing. Ziegler) essentially coincide with those of the RLM. Basically, they reached the following conclusions:

Given the way the war is now developing, we cannot match the production capacity of America and England. In order to continue the war successfully, we must build quality aircraft at the cost of quantity.

The air force cannot afford to lose its air crews, who in addition to their level of training possess extensive front-line experience, because we give them aircraft which are inferior to those of the enemy..."

After these latest briefings, there was unanimity that the two-man cockpit was fully sufficient for the future home-based night fighter role. At the same time the dorsal and ventral barbettes were dropped. After further consideration, it was determined that the high-altitude version (same fuselage with a wing with 50% greater span and 30% greater thickness) would require larger propellers and the DB 614 high-altitude engine. Armor was to be included in the design work at the earliest possible date. Fuel tanks were to be arranged so as to provide maximum possible protection. Speed brakes were required to match the speed of enemy bombers during an attack.

Construction of Prototypes

Cooperation between the front-line units and the manufacturer had resulted in an ideal night fighter aircraft, even though thoughts of heavy fighter and fast bomber versions still tantalized the RLM and Heinkel. At the end of the year Siegfried Günther, now head of Heinkel's design bureau, produced two designs:

The He 219 was now laid out as a two-seat, twin-engined night fighter which was to be equipped with DB 603 engines. Two MG 151/20 cannon were to be housed in the wing roots, with four similar weapons in a fixed, forward-firing installation in a ventral tray. The question of rearwards defense was still not settled, for experience with the He 177 had shown that the FD 131 Z twin-gun barbette, which was also planned for the He 219, was not yet service ready. It was hoped that an MG 131 Z operated by the radio/radar operator would suffice. The nosewheel undercarriage had been reintroduced. The undercarriage members would turn through 90 degrees before retracting into the engine nacelles or forward fuselage. Preliminary steps were made for the installation of an airborne radar in the nose of the aircraft—preliminary, because Telefunken, the producer of the FuG 212, did not expect series production of the C-1 Lichtenstein to start before the beginning of 1943. Ejection seats were planned for crew safety.

The second design powered by the DB 614 (essentially a DB 603 with a three-stage supercharger) was the high-altitude reconnaissance version. In its case, no remotely-controlled weapons stations or wing-mounted weapons were planned

Projected initial production version of the He 219.

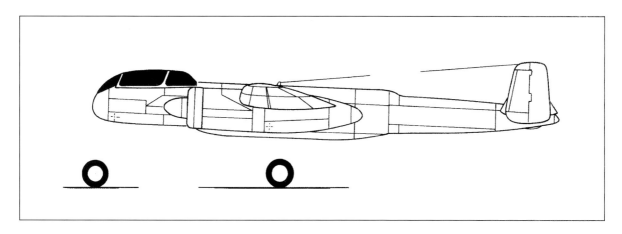

19

The He 219 V1 prototype. Notable features include the stepped fuselage, which was supposed to accept defensive barbettes, four-blade propellers, and the absence of airborne radar.

from the very beginning. Plans included two ETC 1000 bomb racks beneath the weapons tray or four ETC 500s for the fast bomber role. The weapons tray was to house two fixed, rearwards-firing MG 17 machine-guns. Because of the increased all-up weight, the main undercarriage elements were again to mount twin wheels and would be retracted without swiveling through 90 degrees.

Taking into consideration unavailable components, Udet called for a first-class production version of the He 219 powered by the DB 603 and a high-altitude fighter version equipped with the DB 614 and high-aspect-ratio wing. To avoid use of the troublesome remote gun control system, on the fast bomber version the dorsal and ventral barbettes would be moved farther forward, which would make possible the use of a simple hydraulic gun control system. The original requirement for side-by-side seating was dropped.

Based on current experience, the pilot received guidance from his radio/radar operator and finally spotted the bomber visually from a range of about 200 meters. At the time of project planning the observer still had to indicate the direction of the enemy aircraft to the pilot. He accomplished

this with the aid of the Lichtenstein set, which had cathode ray tubes (radar displays) for altitude, azimuth, and distance, and information provided by ground stations. For operation of the FuG Lichtenstein, it made no difference whether the operator sat beside or behind the pilot. A third crew member was not considered necessary for a home defense fighter, as his only task would have been operation of a defensive weapon. On the other hand, a third crew member was indispensable for a long-range night fighter, as he would be responsible for operating the dorsal and ventral gun turrets and maintaining radio communications with home base.

The RLM acknowledged the use of a nosewheel undercarriage as a definite advantage for the night fighter version.

On 8 January 1942 the GL/C decided that a production contract should be issued for the He 219. Following Günther's return, construction began on the first prototype, the He 219 V1. One month later the necessary drawings were eighty percent complete.

On 22 January 1942 *General* Kammhuber and Herr Beauvais inspected the He 219 mockup and acknowledged the arrangement of the cockpit, weapons, and dinghy to be essentially correct. The desired airborne radar was the FuG Lichtenstein, together with an automatic direction finder with autopilot setting and the UHU device (FuG 135 data transmission device). Standard radio equipment was to consist of two FuG 16 ZY sets, plus one FuG 10 P. This allowed frequencies to be selected for Y-control, air traffic control, the *Gruppe* command channel, and the fighter control channel (*Reichsjägerwelle*). By this stage in the project further details were laid down in the RLM. These included the possible use of a GM-1 system for increased performance, braking propellers for reducing speed while closing with a target at night (use of VDM or Messerschmitt reverse-pitch propellers). Armoring of the airframe and engines against 12.7-mm caliber shells up to an angle of 10° from in front for the night fighter and heavy fighter versions, as well as armoring of the engines and cockpit against fire from behind for the heavy fighter and steel spars to protect the engines. The RLM's delivery program determined that the DB 603 C engines would not become available until 1 February 1943. In clear language, this meant that there were no engines available for the He 219 V1 to V5 which could accept the Messerschmitt braking propellers. Delivery of the DB 614 was scheduled for 1 April 1943 and was to affect the He 219 V9. Engines would be taken from the pre-production (Zero) series.

Completion of the night fighter cockpit mockup was requested by 10 March 1942. This cockpit was built by Heinkel as a final mockup.

Since the radiator arrangement was still a matter of contention, the RLM proposed inspecting a captured Russian machine, an Il 2 close-support aircraft, at the DVL. This aircraft had a very well protected radiator installation, which was examined closely by the Heinkel representatives.

On 11 March 1942 Dr. Heinkel wrote to the Daimler Benz AG requesting delivery of the first four DB 603 C engines by the beginning of August 1942. At the same time, Heinkel asked that Daimler Benz director Nallinger take part in the inspection of the power plant mockup in the Marienehe factory in mid-April 1942.

The schedule for further engine deliveries is contained in Table 10.

On 27 March 1942 Director Nallinger wrote to Prof. Heinkel, informing him that delivery of the DB 603 C engines would not be possible prior to October 1942. He proposed that the DB 603 A or B engine be used instead, although these would require smaller diameter annular radiators and associated ventral radiators.

They had not been asleep in England. The British intelligence service operated quickly and efficiently. Production of a true high-performance night fighter by the Germans could threaten the plans of Air Vice Marshall Harris and his Bomber Command. Faced with the prospect of the He 219, as well as the He 280 turbojet fighter and the He 177 long-range bomber, whose potential was overestimated (and not just by the RAF), the British pushed for action. Bomber Command received orders to destroy Rostock and the Heinkel factories. Harris scheduled a series of attacks between 24 and 27 April. The Heinkel factories in the south of Rostock were assigned to the RAF's 5 Group. Rostock was bombed for four nights, and each raid was deadlier than its predecessor, even though the factories and airfield sustained little damage the first night.

Many of the design drawings were destroyed by fire in the third attack. The most effective raid was the last, on the night of 27 April. Much of the modern production facilities went up in smoke. Luckily, the area housing the design bureaus, the model workshop, and the He 219 V1 prototype,

Engines for	Number	EHF Deadline	RLM Deadline
V1	2	01/09/42	01/10/42
Replacement	2	01/10/42	15/10/42
V2	2	15/10/42	01/11/42
V3	2	15/11/42	15/11/42
Replacement	2	01/12/42	01/12/42
V4	2	15/12/42	15/12/42
V5	2	01/01/43	01/01/43
Replacement	2	15/01/43	15/01/43
V6	2	01/02/43	01/02/43
V7	2	15/02/43	15/02/43
Replacement	2	01/03/43	01/03/43
V8	2	15/03/43	15/03/43

Table 10

which was then under production, were undamaged.

As a result of these attacks, part of the production line plus the development and design department moved to the south of the Reich, to Heidfeld air base near Vienna, 680 kilometers away from Rostock. Beginning immediately, the Heidfeld base was referred to as the Schwechat works in all of the Ernst Heinkel Flugzeugwerke's papers and in correspondence with the RLM.

General Kammhuber had had Heinkel submit to him the planning documents for the He 219 with performance diagrams and calculations, which were the result of intensive wind tunnel measurements. He demanded a swiveling seat for the observer, so that he could better operate a defensive position if one were installed and, by rotating his seat 180 degrees, also operate the airborne radar. This soon produced serious problems, however, as it conflicted with another demand for ejection seats. Inevitably, the question of a three-man cockpit arose again. Given the fact that construction of a pre-production series could possibly begin soon, at a conference Kammhuber requested that the formation of a complete operational *Gruppe* of twenty to thirty He 219s begin in April 1943.

On 27 May 1942 the development heads of the Ernst Heinkel Flugzeugwerke

were informed that development of the DB 603 C engine had been halted. Serious crankshaft vibration at a speed of 2,230 rpm and the non-usability of the standard spur gearing meant that the cost of continuing development could no longer be justified. Faced with the growing problem of procuring engines for the He 219 and the He 274, the Heinkel Company reached the conclusion that the DB 603 B with a 300-mm drive shaft, which could be lengthened to 370 mm, and a reduction ratio of 2.07 : 1 could serve as the future power plant for the He 219.

On 14 June 1942 another conference was chaired by *Generalfeldmarschall* Erhard Milch, Udet's successor. The new chief of air armaments saw the He 219 as a replacement for the almost stillborn Ar 240 and declared that the Heinkel met most of the requirements for a night fighter. But to Kammhuber's astonishment, at the conclusion of the presentation he called for a detailed requirement for a Bf 110 replacement. Only afterwards would it be possible to say whether the He 219 could meet this requirement. Even as a successor, it would not be available before 1945.

The *Luftwaffe* command staff immediately raised sharp protests. Kammhuber stated that a successful defense of the homeland would be impossible without a

As Udet's successor, General-feldmarschall Erhard Milch played a fateful role in the struggle to have the He 219 adopted as a service aircraft. Here he is seen in conference with his advisory staff.

high-performance fighter. At this time the successor to the Ju 88, the Ju 188, was brought into the discussion, even though no heavy fighter version was planned. Subsequently, the mock-up of the Ju 188 R night fighter version with Lichtenstein C1 radar was not completed until the end of 1942-early 1943.

At Kammhuber's request Junkers, Messerschmitt, Tank of Focke-Wulf, and Heinkel submitted new proposals. While all other competitors threw their latest developments into the race, Messerschmitt refused to drop his Bf 110. Probably with the tacit approval of the RLM, he merely proposed an improved variant, the Bf 110 G. All of the new designs were based on the coming FuG 220 Lichtenstein SN-2 airborne intercept radar. It required a larger antenna array, the so-called "Big Antlers" (*grosse Hirschgeweih*). This active night fighter device was seen as the successor to the FuG C1. Telefunken predicted an output of 2.5 kw with a range of from 300 to 4,000 meters. The search angle was 120° horizontally and 100° vertically. Under the most favorable conditions this would allow the night fighter to scan an area 7 km x 5 km x 4.5 km in the direction of flight. A rearward-facing radar was also planned to alert the night fighter to threats approaching from that quarter.

The Ju 188 bomber, seen here in its E-version, was a contemporary of the He 219 and was constantly promoted by the RLM for the night fighter role.

A Ju 88 G night fighter equipped with FuG 220 SN2 radar.

Series Production Planning

On 25 June 1942 the RLM advised that the He 219 was now earmarked for large-scale production, 200 machines per month, beginning in 1943. A zero (pre-production) series of 20 machines was to be completed by April 1943. In the planning for this beginning and the subsequent large-scale production of the He 219, the Heinkel Company proposed the following:

a) Zero series production at Marienehe

b) Series production at Marienehe and Mielec

The RLM's planning department had other ideas, however:

a) The Mielec factory is earmarked for other tasks.

b) Construction of the He 219 at Marienehe is undesirable on account of the threat of air attack.

c) The Zero series is to be built at Schwechat. Quantity production in a license factory is still to be determined.

Solving the question of a power plant for the He 219 remained a constant problem. The DB 603 B, which was supposed to power the A-series, ran into considerable difficulties. Heinkel, therefore, turned to the DB 603 A and the Jumo 213. The DB 603 G was supposed to become available in mid-1943, however, it turned out that Daimler Benz never got a hand on the problems with the exhaust-driven supercharger,

and it was not until January 1945 that the He 219 A-7 became the first version to be powered by this engine. Delivery dates for the various engines were discussed on 6 July 1941. Afterwards, the RLM immediately got in touch with Daimler Benz and Jumo, and agreements were reached on delivery dates (see Table 11).

The DB 603 engines were to be delivered immediately in modernized condition with the then standardized running time of 50 hours.

An RLM conference was held on 18 August 1942 at which the Fw 187 *Falke*, a twin-engined *Zerstörer* developed by Focke Wulf before the war, was dropped once and for all. Nine examples of the Fw 187 had been built and flown. The production resources which were released as a result were immediately assigned to production of the Me 210 and Ju 188.

On 31 August 1942 Prof. Heinkel handed over the final drawings for the He 219 to the GL/C. Proposed armament was six MG 151/20 cannon—four in a ventral tray and two in the wing roots. An MG 131 was supposed to be installed for rearward defense. The selected power plant was the DB 603 A, whose maximum output was 1,750 hp.

On 1 September, just one day later, an RLM committee announced production plans. Twelve prototypes were planned,

Type	Quantity	Date	Remarks
Mock-ups:			
DB 603	1	15/07/42	Functioning engine, certified for 20 hrs.
Used as mock-up.			
Jumo 213	1	15/08/42	Mock-up for installation inspection
Jumo 213	1	01/01/42	For mounting in the V10
Engines:			
DB 603A	2	01/08/42	V1
DB 603 A	2	15/08/42	V2
DB 603A	2	01/09/42	replacement for V1
DB 603A	2	15/09/42	V3
DB 603A	4	15/10/42	V4 and replacements for V2 and V3
DB 603A	2	01/11/42	V5
DB 603A	3	15/11/42	V6 and replacement for V4
DB 603A	3	01/12/42	V7 and replacement for V5
DB 603A	2	15/12/42	V8
DB 603A	4	01/01/43	V9 and replacements for V6 and V7
DB 603A	2	15/01/43	replacements for V8 and V9
Jumo 213	2	15/01/43	V10
Jumo 213	2	01/02/43	replacements for V10

Table 11

followed by the manufacture of 173 examples of the A-0 pre-production variant between March 1943 and September 1944. Production of a further 117 aircraft would be possible from January 1944 if Special Commission F3 (Heinkel controlling body) made available materials for further machines. Prospective production sites were located in Budzyn near Krasnik and Mielec, towns in the *Generalgouvernement* (occupied Poland). These factories had been inspected some time earlier by the Reich Minister for Armaments and War Production. A request was now made for a priority rating, which was needed to speed up production. Capacity information was accompanied by a list of planned manufacturing times and production costs (see Table 12). Since the material costs listed therein do not include the cost of equipment provided by the Reich, such as weapons, power plants, propellers, and radios, the total average cost was rather higher (see Table 13).

By September 1942 construction of the He 219 V1 was nearing its final phase. Prof. Heinkel issued the following statement about the machine's empennage:

"Today I saw and also shook the twin tail mounted on the He 219. I consider the

Machines	Production Hours	Material Costs in DM
V1 – V4	110 000	210 000
1 – 10	86 000	180 000
11 – 20	76 000	175 000
21 – 30	66 000	170 000
31 – 40	60 000	165 000
41 – 50	56 000	160 000
51 – 60	52 000	160 000
61 – 70	48 000	155 000
71 – 80	45 000	155 000
81 – 90	42 000	150 000
91 – 100	40 000	150 000
above 100	38 000	150 000

Table 12

Machines	Average Price in DM
V1 – V4	558 000
V1 – V20	428 000
V21 – V40	367 000
V41 – V60	334 000
V61 – V110	228 000

Table 13

fuselage to be too weak and have doubts that we can rely on the twin tail alone. Therefore, contrary to our agreement of 3 August 1942, Point 5 of the statement of 4 August No. 270/42, design work on the central fin and rudder, which was begun at the beginning of August, must be pursued and completed as quickly as possible. As a result of the design we will eventually be able to fit a new rear fuselage to the V2 with the new central fin and rudder."

After the He 219 received the DE priority level, the Heinkel Company had no more difficulties obtaining aircraft parts from the subcontractors. The only serious delay involved the VFM undercarriage; these components for the He 219 V1 were not delivered until the second half of September 1942.

Design of the undercarriage was finalized with the V1, and it subsequently remained unchanged apart from some local strengthening. Each element of the main undercarriage was equipped with two double-brake 840 x 300 wheels and was retracted rearwards into the engine nacelle hydraulically via folding struts and accumulators. The oleos were fitted with compressed air shock absorbers with grease damping. The nosewheel leg was equipped with a 770 x 270 wheel without brake. It had a steering damper whose purpose was to keep the wheel fork in a straight position. During the course of trials additional oil damping was tested but was not adopted. The oleo was equipped with a compressed air shock absorber with grease damping. The nosewheel member was retracted hydraulically by way of folding struts and accumulators. During the retraction sequence the nosewheel turned through 90°. This was achieved by an installation in the top of the wheel fork. During retraction, a spring-loaded lever with

Retraction cycle of the He 219's main undercarriage.

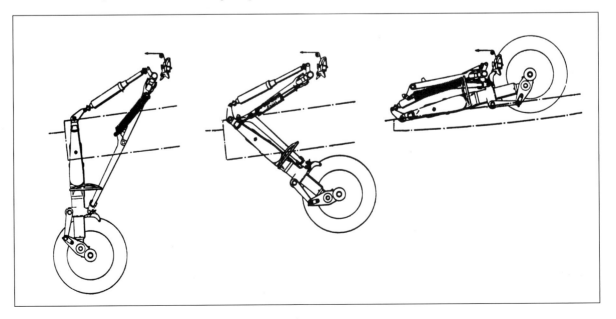

26

roller, which limited deflection to 60° by means of a hook and stop, contacted a guide and was folded into the fuselage while turning through 90°. When the undercarriage was lowered, the steering damper ensured that the nosewheel returned to the straight position.

In October 1942 the He 219 was second on Heinkel's development priority list behind the He 177, and, in the period that followed, production capacity for the He 219 constantly lagged behind the planning. Meanwhile, construction of the He 219 had reached the stage where the question of a location for testing had to be discussed. Since Schwechat's runway was still not completed, the He 219 was to be test flown at Marienehe. Blast pens were built as a security measure. Important components were to be dispersed in order to avoid a total loss in the event of air attack.

Another important meeting took place in Berlin on 4 September. *Generalmajor* Vohwald, *Generalingenieur* Hertel, and *General* Kammhuber compared performance diagrams for the Me 210, Ju 188 N and G (both projected variants), Do 217 J, and Ju 88 C with those of the He 219 A. The comparison of these calculations and of raw material requirements revealed clear advantages for the Heinkel design.

The aircraft's engines were to be equipped with flame dampers for the night fighter role. At night the pale blue exhaust flames could be seen from a great distance and had to be effectively covered or even eliminated. Flame dampers fitted to the engines of the Ju 288 had produced good results, and their use on the He 219 was investigated.

In armament planning, a mockup installation of four MK 108 or 103 cannon in a fuselage tray was tested, for it was desired to experiment with a much heavier armament with as many MK 108s as possible on one of the first prototypes.

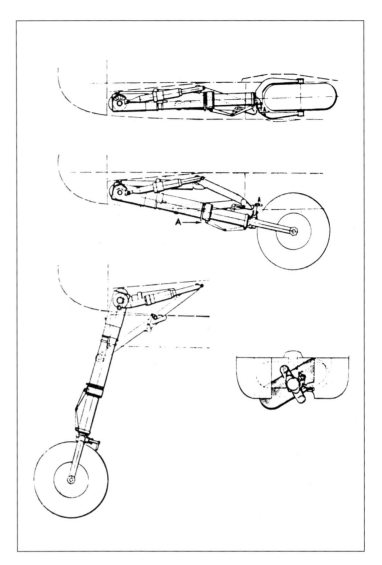

On 30 September it was announced that the central fin and rudder would be completed by 30 November 1942, however, at the beginning of November Prof. Heinkel ordered the work stopped, even though the completion deadline still stood and the first tail unit was supposed to enter service at the end of January 1943. Obviously, he was now convinced of the structural strength of the twin tail unit.

The RLM placed the He 219 in Delivery Program 222, however, the planned production figures could not be met on account of the extraordinarily large number of men leaving the German labor force. In a letter to the RLM, the Heinkel Company

Retraction cycle of the He 219's nosewheel.

made reference to the type's final mockup inspection, which had not taken place until 21 April 1942, and predicted that the prototype would make its maiden flight at the beginning of November. Completion of another ten prototypes was expected by April 1943. Quantity production could begin in January 1944 at the earliest.

Walk around the He 219 Prototype

The He 219 V1, W.Nr. 219001, at Rostock Mariene-he. By the time these photographs were taken the fuselage lines had been revised to eliminate the steps for the gun barbettes, and it bears the radio call sign VG + LW, which means that they were taken after January 1943. The machine does, however, still wear the then standard camouflage finish of RLM 71/65.

Front view of VG + LW.

Retracted entry ladder and self-closing footsteps, which provided access to the cockpit. The hole above the rear of the ladder door gave ground personnel access to the locking/unlocking mechanism for the ladder (above left).

View of the cockpit area from the port side of the aircraft. Note the opened panels revealing the control linkages. The entry ladder is in the retracted position. This machine has been fitted with the nose fairing incorporating mountings for the FuG 212 antennas and has been painted matte black (above center).

Instrument panel and internal windscreen, with armored visor folded forward. The reflector gunsight is a Revi 16 B/G (right).

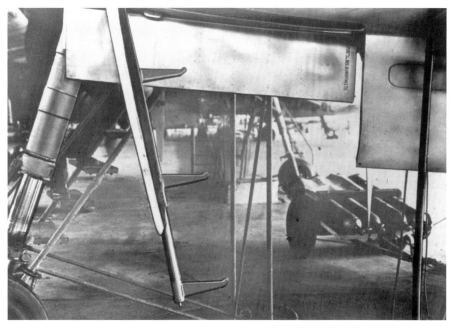

Entry ladder in the extended position. Prior to retraction the bottom step was pushed up to the second step and locked (above right).

Armored visor in the raised position in front of the armor glass internal windscreen. On Mosquito-hunter versions this armor was removed as a weight-saving measure (left).

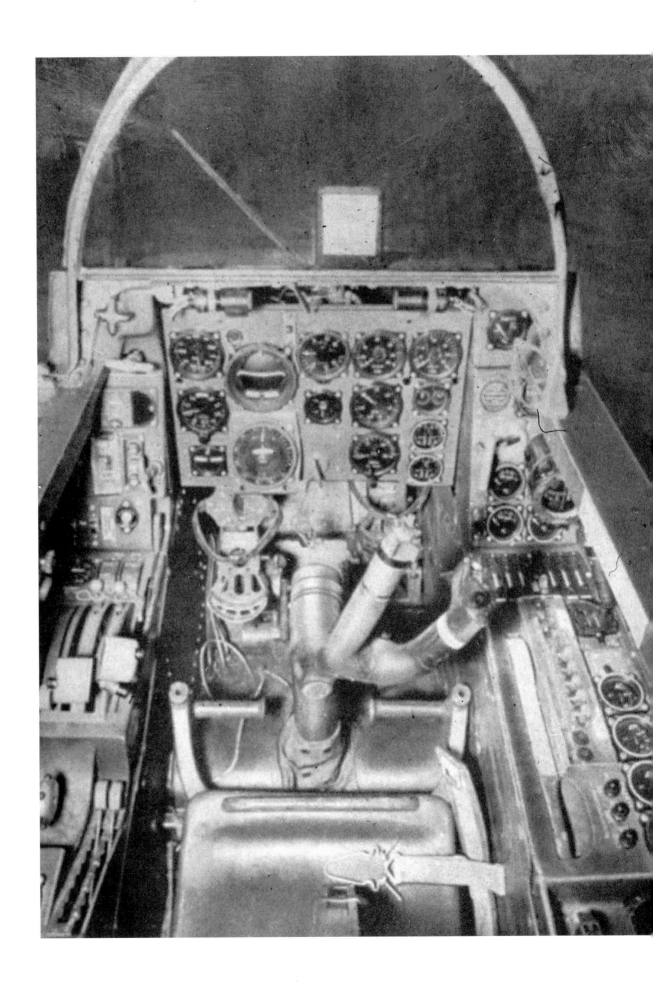

Pilot's position in the He 219. All instruments and control devices were placed within clear view and easy reach of the pilot, one of the innovatory features of the He 219's design (far left).

Pilot's starboard console. Visible are the three fuse boxes, the oxygen supply, instrument console with starter and auxiliary starter switches, fuel gauges, and fuel warning lamps (left).

Ejection seat rail mountings; both seats have been removed.

The indicator light panel (12) on the port console of the pilot's cockpit was used to monitor the position of the undercarriage and landing flaps. A green light indicated that a component of the flaps or undercarriage was retracted, and a red light that it was down and locked.

This jettisonable canopy was fitted to all two-seat versions of the He 219 (above left).

The accumulator provided the necessary pressure-oil reserve for the aircraft's eight wheel brakes. It was located directly between frame 9 and the first fuselage fuel tank. Even with the accumulator out of action, wheel landings could still be carried out safely provided that there was still circulatory pressure available and at least one engine was running (above left).

Located under the first fuselage cover panel behind the radar operator's position was the first of three fuselage fuel tanks with a capacity of 1,100 liters of B4 aviation gasoline (right).

The middle of the three fuselage tanks held 500 liters of fuel and, like all of the aircraft's fuel tanks, was self-sealing. Photo taken looking forward (far right).

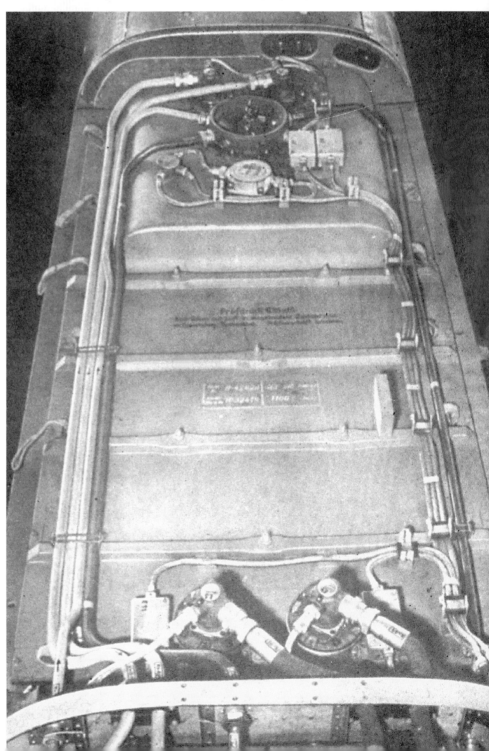

Photograph of the forward cockpit armor of the He 219 A-012 (above left).

Separation point at frame 9a. The entire cockpit assembly was bolted to this frame (below left).

He 219 prototype construction showing wing and fuselage assembly (above).

This entry hatch was located in the rear fuselage of all He 219s. It provided access for servicing. Later it was used to gain access to the "schräge Musik" mounting. Photograph taken looking forward (above center).

Oxygen for the crew was contained in these ball-shaped bottles, which were mounted on the roof of the fuselage above the entry hatch (bottom center).

 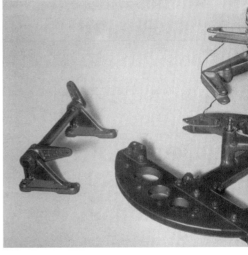

The supplementary heaters, known as "Kärcher Stoves," presented the Heinkel engineers with some problems during the test phase. Here one such heater is seen in the rear fuselage. Failure of these devices was included in a summary of causes for He 219 crashes (above left).

The tailplane control unit, with actuating lever and attached control rod, was located directly in front of the rear fuselage heater (center left).

Blast tube and mounting for the MG 151/20 mounted in the starboard wing root.

Ventral weapons tray of an He 219 A-0/R3. The open access panels provide a view of the MK 103 installation. At the top of the photo, the ejector for the port wing-mounted MG 151/20 on the underside of the wing (above right).

Components of the aileron control quadrant (above center).

View of the installed aileron control quadrant through an open access panel (above center).

Access to the elevator locking device and coupling shaft was provided by an access panel on top of the aft fuselage (left).

Upper engine servicing panels closed (above).

Upper engine servicing panels opened. Note the warning *"Achtung! Verbrühungsgefahr"* ("Attention! Danger of Scalding") on the liquid coolant tank (center).

Cooling gills in the open position seen from beneath the starboard engine nacelle. Note the mounting for the flame damper tube (bottom).

Main undercarriage oleo with wheels removed. Note the position of the brake line and brake hoses.

View of the starboard side of the engine with cooling gills open and a segment of the annular radiator visible (left).

This servo motor set the position of the cooling gills by moving a ring attached to the gills (above).

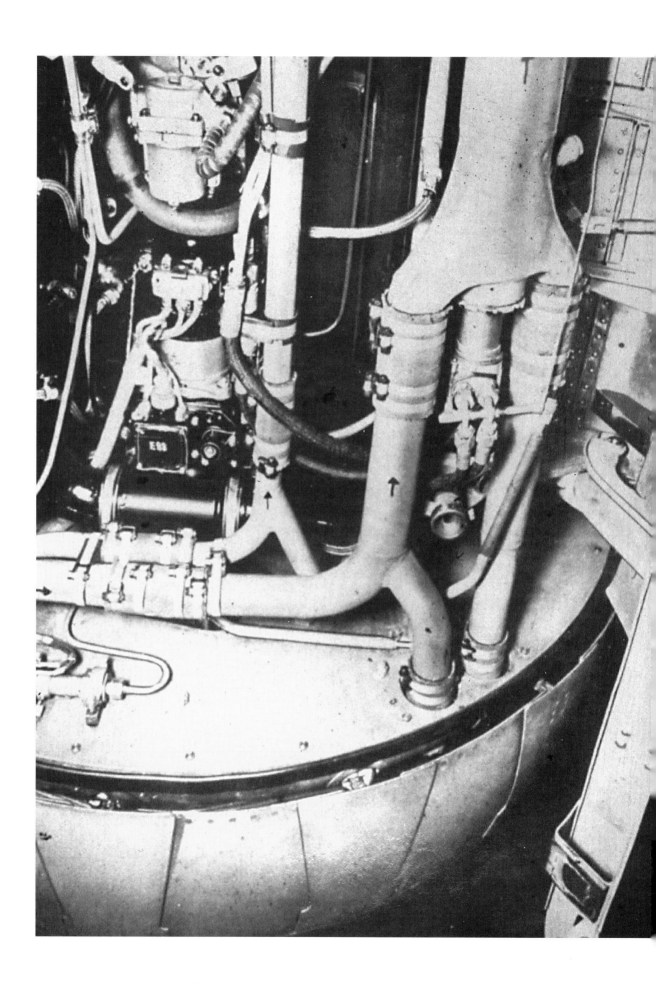

Removal of the access panel on the bottom of the cowling gave access to the DB 603 A engine. The large hoses were part of the coolant return system (far left).

The starboard main undercarriage of the He 219 (center).

This photograph offers a good view of the main undercarriage and wheel well with undercarriage attachment points (left).

View of the port wheel well with landing gear doors removed (bottom).

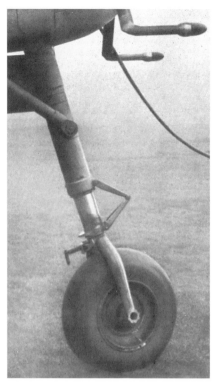

Main undercarriage release servo mounted on the main spar in the wheel well (top left).

Retraction cylinder and main undercarriage attachment point (top center).

The two hydraulic fluid containers (85 l each) were housed in the wheel wells and riveted to the wing structure (left center).

The nosewheel of an He 219 A-0. Note the FuG 220 antenna mounts (top right).

The nosewheel well seen looking forward. The nosewheel leg has not yet been installed (top center right).

Nosewheel well and nose-wheel door seen looking aft (large photo).

Accumulator with attached hydraulic pressure gauge for the nosewheel undercarriage (left bottom).

Testing

On 6 November 1942 the He 219 V1 rolled out of the final assembly hall at Marienehe for its first flight. At 1308 hours it lifted off from the runway with company test pilot Peter at the controls.

Peter's overall impression of the machine was good. After landing at 1318 hours he raised the following points:

1. Elevator balance very positive, forces light, but no change necessary.
2. Rudder forces and effectiveness very good.
3. Aileron forces are very high with the flaps in takeoff position. The aircraft was not flown with the flaps in other positions. Aileron trim proved ineffective at speeds of 200 to 280 km/h. It was as if the ailerons were blocked during the entire flight.
4. There was no light indication when the nosewheel was retracted and the hydraulics did not shut off. When lowered, the nosewheel positioned itself at an angle of 45° and remained there in spite of repeated retraction and extension. Landing with the nosewheel at an angle produced heavy cockpit vibration.
5. Power plants worked well, as evidenced by exhaust temperature of 95° and oil inlet temperature of 65°. Propeller pitch control functioned correctly.
6. Twin rpm indicators led to errors.
7. Radio systems must be checked.
8. Throttle regulation is very hesitant and the motion is not rectilinear. It is very difficult to set the engines to a desired speed.

In a letter to *Generalfeldmarschall* Milch dated 7 November 1942, Dr. Heinkel pointed out that the He 219 V1 had completed its maiden flight ten days before the

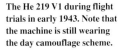
The He 219 V1 during flight trials in early 1943. Note that the machine is still wearing the day camouflage scheme.

promised deadline. He went on to state that no major problems had arisen.

During the third company test flight on 9 November 1942 the He 219 V1 was involved in a minor accident. The crew were not injured, and the aircraft sustained only minor damage. Pilot Peter landed the machine with too much airspeed in conditions of reduced visibility caused by a heavy rain shower. As a result, the aircraft touched down at the end of the first third of the runway. The runway surface was slippery on account of camouflage paint applied to the runway, and Peter was unable to bring the aircraft to a stop. It rolled off the end of the runway at a speed of about 20 to 30 km/h and the nosewheel sank into the soft ground. As a result, the nosewheel leg broke off and the machine tipped forward onto its nose. It would be ready to fly again by 15 December 1942.

On 14 November 1942 a conference was held by Generalfeldmarschall Milch; among the topics was the He 219 delivery situation. It was determined that the original large-scale production plans contained in Delivery Program 222 had been based on an error, and Milch authorized deliveries based on the proposals made by Ernst Heinkel Flugzeugwerke with quantity production to begin in 1944.

The He 219 V1 made its fifth flight on 17 November 1942 and its sixth on the

18th. The most serious complaint so far was of inadequate longitudinal stability. Some rudder vibration had also been observed, and this was traced to airflow effects from the engine nacelles. On 19 November 1942 the He 219 V1 was demonstrated to *General* Kammhuber at Rostock-Marienehe. At that time the He 219 V2, V3, and V4 were in the final assembly stage at Vienna-Schwechat. In the meantime, considerable difficulties had arisen from the request for 200 specialists for production of Heinkel aircraft, a problem which could only be temporarily bridged at the cost of He 219 production. In the month of October 1942, alone, 186 specialists had been lost to the He 219 production program as a result of being called up. Based on a ten-hour work day, this resulted in the loss of 55,800 man hours by 20 November 1942.

On 20 November 1942 the He 219 V1 completed its seventh flight. It was noted that static stability was noticeably better than on the previous flight. The reason given was a modification to the outer landing flaps. Tests with smoke cartridges had shown that airflow was still good on the engine nacelles, but was being disturbed just beneath the horizontal stabilizer and next to the lower part of the vertical stabilizers.

On 24 November 1942 the He 219 V1 completed its tenth flight in the hands of

The following series of photographs was taken during a flight demonstration by the He 219 V1 during flight trials. The aircraft is now finished in an overall matte black night finish (above and following eight photographs).

pilot Franke. Franke's assessment of the machine was largely in agreement with that of pilot Peter.

By 26 November the He 219 V1 had completed a total of eleven flights, the results of which are summarized below:

1. No difficulties arose with the power plants, the engines proved extremely trouble-free.
2. Overall, the taxiing characteristics of the nosewheel undercarriage were seen as good, however the nosewheel places itself at an angle if turned on the spot and subsequently refuses to return to the neutral position.
3. Takeoff and landing was extraordinarily simple, and in no case was there any bounce on landing.
4. Directional stability was adequate with center of gravity in the mid-position.
5. Static stability about the normal axis was too weak, an investigation is underway with enlarged vertical stabilizers.
6. Rudder forces were too weak in the neutral position, too high beyond 30 percent deflection. Irregular rudder vibrations occurred, probably on account of airflow breakaway around the engine nacelles.
7. Control forces were good in the neutral position, too high beyond 30 percent deflection.
8. At cruising power the aileron forces were good up to 40 percent deflection, beyond that too high.
9. Undercarriage retraction and extension is very quick. Initially, there was a problem when retracting the undercarriage, in that the nosewheel was turned by the airflow and could not be retracted. This problem was solved through the use of a stronger return spring.
10. In high speed flight the landing flaps remain extended 5°.
11. Maximum airspeed achieved 480 km/h.
12. Maximum boost altitude of 6 000 meters was not reached, drop in boost pressure beginning at 5 000 meters.

Flights 13 to 16 were carried out on 2 December 1942, and the machine was subsequently taken to the workshop for modi-

fications. The fuselage of the He 219 V1 was lengthened by 940 mm and was fitted with 30 percent thicker skinning. The upper and lower lines of the fuselage were smoothed, resulting in the elimination of the steps originally intended to accommodate the remotely-controlled barbettes. It was hoped that lengthening the fuselage would result in an improvement in stability about the longitudinal and normal axes, reduce the trim changes during landing, and eliminate the vibration that had been observed. In the course of this conversion the V1 retained the original, smaller fins and rudders. As well, inaccuracies in the construction of the ailerons were rectified, and trim balance was achieved by installing counterbalances.

The planned flight test program included a performance flight by Herr Neidhardt of the *E-Stelle*, an evaluation flight by *Hauptmann* Streib, investigation of single-engine flight characteristics, and wool tuft tests to check airflow conditions.

By 15 January 1943 the He 219 V1 had completed 46 flights with a total duration of 30 hours and 40 minutes.

The He 219 V2 was first flown on 10 January 1943, and by 15 January it had logged approximately 1.5 hours in six flights. The aircraft was test flown by *Oberstleutnant* Petersen, *Major* Daser, *Fl.-Stabsing.* Böttcher, *Fl.-Stabsing.* Beauvais, *Fl.-Stabsing.* Neidhardt, and *Fl.-Stabsing.* Bader. On 15 January 1943 *Major* Streib of *XII. Fliegerkorps* test flew the He 219 V2. Results of these test flights largely confirmed the assessments made by the Heinkel Company's pilots and raised the following points:

a) Performance was poorer than guaranteed. The *E-Stelle Rechlin* clocked a speed of 475 km/h at low level compared to a promised speed of 490 km/h.

b) There was still unexplained vibration in the tail section.

c) Longitudinal stability was weak, especially when the landing flaps were down.

d) Excessive changes in trim while retracting the landing flaps and about the longitudinal and normal axes during speed and power setting changes.

e) Overall, control forces were too great.

Meanwhile, the rake of the nosewheel leg was increased and the main undercarriage legs were moved forward by 220 mm, resulting in improved taxiing and landing characteristics. Other important experiments were conducted at this time:

a) Lengthening of the engine nacelles.

b) Doubling in size of the vertical stabilizers and rudders.

c) Modification of the aileron balancing and installation of a 100% mass balance.

d) Modification of the landing flap retraction arrangement.

Since the experiments aimed at eliminating vibration and increasing longitudinal stability had failed to produce satisfactory results, as already mentioned, the fuselage of the He 219 V1 was lengthened, and it was anticipated that the same would be done to the V2.

On 28 January 1943 *General* Kammhuber revealed that he had spoken to the Reichsmarschall about the He 219 and that the project had received top priority. Furthermore, it was planned to increase monthly output to 50 aircraft by the end of 1944 and to 100 aircraft per month during 1945, provided that the promised performance figures and handling characteristics were achieved. Operational testing was to be conducted by I./NJG 1, a unit of *XII. Fliegerkorps* (*Major* Streib).

On 2 February 1943 Prof. Heinkel was able to inform *General* Kammhuber of the successful conversion of the He 219 with the now lengthened fuselage and contra-rotating motors. The modified He 219 V1 re-

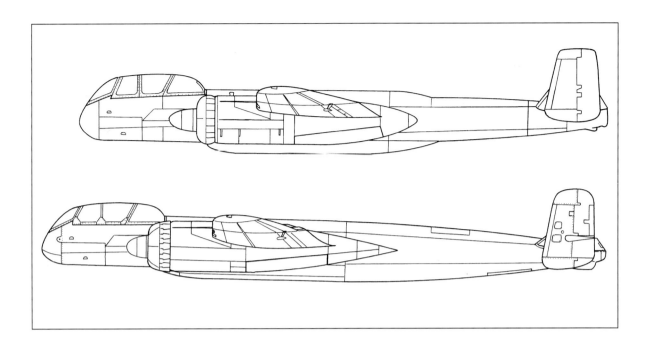

turned to the air on 30 January 1943 and soon demonstrated the desired handling characteristics. In mid-January at Peenemünde an armament of four MK 108 cannon was installed in the He 219 V1. Meanwhile, the He 219 development section had been moved from Rostock to Schwechat.

Dr. Heinkel had promised the staff a bonus of RM 3,000 for successfully completing the conversion of the He 219 V1 ahead of schedule (8 February instead of the planned 10 February 1943) and a further RM 1000 for each additional day, which meant that he was prepared to pay out RM 10,000. A trustee of the Ernst Heinkel Flugzeugwerke forbade him to do so, however, as he related in a letter to Meister Richmann in the Marienehe factory dated 19 February 1943.

Concerning the state of He 219 testing, the following points were recorded in the minutes of meetings held on 10 and 15 February 1943:

a) Conversion of the He 219 V2 to lengthened fuselage to provide an additional aircraft to the He 219 V1. The fuselage was to be streamlined and the machine was to receive new vertical and horizontal tail surfaces, new ailerons, and lengthened engine nacelles.

b) The lengthened fuselage was to be adopted for all aircraft beginning with the He 219 V7 (previously planned to begin with the O-series).

c) All V-machines (prototypes) were to be fitted with the larger tail surfaces.

On 16 February the folding section of the V2's canopy became caught in a crosswind while the aircraft was taxiing. The canopy broke free and was struck by the starboard propeller, resulting in major damage to the fuselage. Immediate steps were taken to prevent a recurrence of this incident.

On 18 February 1942 the testing schedule for the various prototypes was as follows:

He 219 V1:

Flight performance and handling characteristics.

He 219 V2:

Diving performance and handling characteristics with the DB 603 V (from Messerschmitt).

He 219 V3:

Power plant tests with the DB 603 A, cockpit heating, wing de-icing, tail de-icing, autopilot and compass system (from mid-July 1943).

He 219 V4:

Radio testing, without the FuG 10 P, however (beginning end of February 1943).

He 219 V5:

Testing of fixed armament (beginning March 1943).

He 219 V6:

Aircraft to be completely equipped, including FuG 10 P. After conclusion of radio equipment tests delivery for service trials.

He 219 V7 – V9:

After completion of test flights (beginning April 1943) delivery for service trials.

He 219 V10 and V11:

After completion of test flights (beginning May 1943) delivery for service trials.

He 219 V12:

Test-bed for remotely-controlled barbette (beginning June 1943).

Difficulties with the delivery of engines meant that the program could not be carried out as planned, however. Promised power plants were held back or, as was announced on 25 February 1942, even allocated to the Do 217 production line. Through the intervention of *General* Kammhuber, a promise was made to deliver the He 219 V7 and V8 to *XII. Fliegerkorps* in mid-May 1943.

According to Heinkel, the output of He 219 pre-production machines projected in Delivery Plan 222 could only be met if the company received an additional 620 workers (half German and half foreigners). The breakdown of specialists was 6 percent electricians, 20 percent sheet metal workers, 50 percent fitters, and the rest semi-skilled laborers. The shortfall by 6 March 1942 of 508,000 man-hours was to be made up through overtime. The most important factor in adhering to the schedule, however, was the guaranteed delivery of engines. Heinkel also complained about the release of designer engineers to Junkers for the Ju 352 development program, as well as to Rheinmetall-Borsig. In the opinion of the Ernst Heinkel Flugzeugwerke, Rheinmetall had had the cheek to release the former Heinkel specialists to the *Wehrmacht* and then turn around and request replacements from Heinkel. Following the intervention of *General* Vorwald the men were returned to Heinkel. A decision on quantity production was not expected until the approval of Delivery Plan 223 (mid-March 1943), since the program was required to identify free production capacity within the industry. The Heinkel Company (Dir. Heyn, Rostock factory) was against production of the aircraft at Rostock-Marienehe for the following reasons: "Rostock is in Danger Zones 1 and 1a. For this reason production of the He 111 has been dispersed and Mielec/Budzyn has been activated as the second factory. It would therefore be irresponsible and inexplicable if this design, which has been characterized as extraordinarily important, should be transferred to Rostock. The English would soon discover where the new night fighter is being built and, if it proved a serious enough threat, it would not be long before the Rostock plant was heavily bombed. If production goes ahead at Rostock, the involvement of at least one other company would be vital, while on the other hand the small monthly production figure would make this unprofitable. It is impossible to reach 50 aircraft per month by the end of 1944. Delivery of the first machine before April is out of the question."

The following lines from the above-cited letter reveal the situation of the aircraft industry very clearly: "Director Dr.

Heyn has stated that it would be impossible to pursue development using Russian women and non-German speakers who could neither read drawings nor understand anything else." Elsewhere it said: "The RLM therefore has no cause to be amazed that we are presently behind schedule." Effective immediately, *Oberst* von Lossberg was named commissar in charge of the He 219 program. It was his task to finally solve the matter of series production by consulting with the *C-B Amt*, industry advisors, and all participating sides. On his authority, from April 1943 production of the He 219 was to be decentralized. Fuselages were to be built at Mielec/Budzyn. Me 323 *Gigant* aircraft would then transport the fuselages two at a time to Vienna-Schwechat.

The following is a summary of flying activity by the He 219 prototypes during the reporting period:

- V1 a total of 59 flights, 29.11 hours,
- V2 a total of 32 flights, 13.11 hours,
- V3 a total of 3 flights, 1.34 hours.

The He 219 was now flying with the extended fuselage and contra-rotating engines. The vibration reported earlier had now disappeared. The aircraft were test-flown by the *E-Stelle Rechlin* and *XII. Fliegerkorps.*

He 219 V1:

a) Greater stability about the longitudinal and normal axes has been achieved through the adoption of the extended fuselage and larger tail surfaces.

b) Fuselage and tail vibration has been eliminated as a result of lengthening the fuselage and modifying the tail surfaces.

c) Serious trim changes during lowering of the landing flaps have been overcome through use of elevator trim tabs.

He 219 V2:

a) As a result of installing contra-rotating engines, the machine has good directional and longitudinal stability in spite of the short fuselage. However, a lengthened fuselage and larger vertical stabilizers are required for improved stability at idle. The contra-rotating engines make the machine much smoother and eliminate the sympathetic vibrations.

He 219 V3:

a) At present is being used for tests with the flame dampers, de-icing, and heating systems. Results of these are not yet available.

During trials the V1 and V2 experienced shudder in the main undercarriage. The causes were sought at Rechlin using the brake test bench. Since efforts to reduce aileron control forces had proved unsuccessful, completely redesigned ailerons were fitted to the V2. Aerodynamically boosted ailerons were planned for the V1, and these were also tested. Future machines were to be fitted with revised cockpit glazing, which had a lower profile and a more aerodynamic joining with the fuselage.

During this period *Generalluftzeugmeister* (Minister of Air Armaments) Milch continued his efforts to have the Ju 88 or Ju 188 adopted instead of the He 219. In many other cases, the State Secretary could not bring himself to make a clear decision, which was required by the precarious situation. It is understandable that this man, pressed by Hitler, was more interested in being able to present figures showing high production numbers, for introducing a new and unproved design could only be done at the cost of current production. His clear preference for the Junkers company does, however, allow some curious conclusions to be drawn. But Kammhuber wanted the He 219, from which he expected more. In an effort to put an end to this intriguing, the general ordered I./NJG 1 at Venlo to conduct service trials with the He 219. For this purpose Heinkel equipped the V7 and V8 with the MK 108 armament package. It was hoped that a truly decisive success would be achieved. The RLM now demanded a comparison flight, for the most

serious argument, which was now being raised again, was: why develop a new machine if a slightly-modified version of an existing machine will suffice? Kammhuber, who had followed the He 219 flight test program since the sixth test flight, was certain that it would prevail.

On 25 March 1943 everyone with anything to do with the affair was at Peenemünde. The RLM had chosen *Oberst* Lossberg to pilot the Ju 188, while *Major* Streib was to fly the Heinkel design.

The Ju 188 E-1 which took part in the fly-off was a standard bomber with external bomb racks and defensive armament removed. In spite of this, it had to take part in the trials, because of Milch's efforts to have a night fighter version of the Ju 88/Ju 188/Ju 388 series win out over the He 219. No night fighter version of the Ju 188 was ever built. Even Dr. Ernst Heinkel made an error. In his book *Stürmisches Leben* he also wrote of the Ju 188 as a night fighter. He had probably been misled by the Ju 188's larger fin and rudder and its BMW 801 power plants, and therefore confused it with a Ju 88 G.

The competition was planned to include speed trials and a mock dogfight. Herr Francke wrote the following in a letter to Dr. Heinkel: "Ju 188 arrived this morning. The He 219 took off at 1100 hours piloted by *Major* Streib, the Ju 188 at 1300 hours with Herr Hermann at the controls. The first round ended one to nothing for the 219. Lossberg has taken over the 188. The 219 is more maneuverable and outturned the 188 in a dogfight. A speed advantage of 25 km/h has been confirmed. The Ju 188's theoretically superior rate of climb has not been confirmed."

A number of pilots flew the He 219 V1 during the course of these comparison flights on 25 and 26 March 1943: *Major* Streib, *Major* Daser, YY *Fl.-Stabsing.* Friebel, *Fl.-Stabsing.* Neidhardt, *Fl.-Stabsing.* Böttcher, Hptm. Jope, Hptm. Grossholz, *Flugk.* Wandel, and *Stabsing.* Bader. The overall opinion was that handling characteristics were "in order."

Generalfeldmarschall Milch subsequently inspected the aircraft and was informed of the following results:

a) The performance of the He 219 is at least 25 km/h and at most 40 km/h higher than that of the Ju 188.

b) The handling characteristics of the He 219 are more akin to those of a fighter. Elevator forces were found to be too high.

c) Production-wise, the He 219 is one of the best and most modern aircraft. Production time was estimated as less than that of the He 111 and much less than the Ju 188, approximately 2/3 the time.

d) The He 219 V1 has been flying for four months. Its handling characteristics and performance are very good; it represents a significant advance compared to previous designs. Today's results surpassed performance estimates.

Though impressed by what he had heard, Milch subsequently declared that for supply reasons producing just 50 aircraft per month was not worthwhile. Representative Friebel advised that the RLM was ordering that the type enter series production with a maximum output of fifty examples per month.

Trials by the *E-Stelle Rechlin* with the MK 108 heavy cannon installed in the He 219 V5 resulted in difficulties. Initial at-

tempts resulted in bulges in the weapons tray because the powder gases were not being vented. The shell ejection chutes normally took care of this function, however, there were none in the weapons tray, as the spent casings were supposed to be collected for recycling. Good results were achieved after necessary modifications were carried out. Weapons trials continued until May 1943 using the He 219 V7, V8, and V9 as test-beds. By this time the He 219 V1 to V3 were being flight tested and the V4 to V6 were ready to fly.

The first accident involving an He 219 took place on 19 April 1943. Herr Jenbach received the following telex from the Heinkel factories in Vienna:

Heinkel Vienna to
Heinkel Jenbach
19/4/43, 2059 hours   Telex no. 292
Message for Dr. Heinkel
Subject: accident He 219 V3
At about 1500 today a maintenance test flight was carried out prior to a night flight by the He 219 V3, in which wing damping and engine test equipment had been installed. On landing, pilot Schuck al-

lowed the machine to descend too rapidly. After a very heavy landing, Schuck took off again and prepared for a second landing.

This time he touched down just short of the field railway. The wheels struck the tracks, causing the undercarriage to collapse, and the machine landed on its belly. Damaged were:

1. Both main undercarriage legs torn from the wings,

2. engine nacelles damaged,

3. rear fuselage and tail torn open,

4. propellers bent.

Further damage cannot be ascertained at this time and will be assessed tomorrow after the machine is recovered, once the accident site has been photographed as per standing orders and inspected by the air and construction supervisors.

Our preliminary estimate of repair time is 6,000 to 8,000 hours.

Dir. Franke and Dir. Schaberger were requested to advise by telex as soon as possible this morning which aircraft are supposed to conduct tests with the wing damper and power plants, so that we do not lose any time in testing and can install the appropriate equipment in the designated machines. It is noted that V4, V5, and V6 are flyable, so that one of these machines may be assigned to factory trials. A representative from Rechlin will be here tomorrow to collect the V4 and ferry it to Rechlin. We therefore propose that the V5 or V6 be chosen.

This telex is being sent to:
Dir. Dr. Koehler
Dir. Dr. Franke
Dir. Dr. Schaberger.

One day later the following telex was sent to Prof. Heinkel concerning the V3 accident and the consequences for the pilot:

Heinkel Rostock to
Heinkel Jenbach
20/4/43 1331 hours   Telex   No. 18475

1. From Dir. Francke

Subject: Accident involving the He 219 V3 pilot Schuck.

a) Under no circumstances should a company test pilot like Schuck have such an accident. The next time I am in Vienna I will investigate the accident closely, but I do not believe that he is pardonable.

b) I would order Schuck grounded for three months to begin with. Schuck appears to have lost his nerve.

c) Schuck is certainly not too ambitious to go around—on the contrary, during my last visit to Vienna I was told by various people that Schuck makes a go-around on 50% of all his flights and that his flying leaves an uncertain impression. I subsequently spoke to Schuck, told him that he should not have any false ambition, and that if he was overworked or weary of flying he should tell me, there is no dishonor in that. Schuck then told me that he always preferred going around to making a bad landing.

d) Schuck has had a total of three crashes with us, the present one with the He 219 V3, those with the He 177—ground loop at Schwechat and a less serious ground loop at Rostock.

e) The airfield at Schwechat is not in the condition required for extensive testflying by aircraft like the He 177, He 219, and He 280. Work on the airfield must be ended in the near future. The tracks of the field railway, on which the main undercarriage of the He 219 V3 became hung up, run straight across the runway just before the threshold.

New Escape System

Following repairs the V3 was assigned to Messerschmitt Augsburg for testing of a newly-developed coupled power plant. The V1, V2, and V5 remained with Heinkel, while the V4 and V6 went to *E-Stelle Rechlin*. There test pilot Oblt. Eisenmann was given the task of carrying out flight trials with the new ejection seat escape system. These tests lasted until the summer of 1944. One of the two pre-production He 219s allocated to Rechlin was also used in the trials. Eisenmann first conducted several "test shots" using a test bench at Heinkel Marienehe. He sat on a seat mounted on rails on a sloped firing ramp, activated the release, and was then braked sharply at the end of the ramp. Everything was filmed, and in the beginning every test resulted in a complaint that the great acceleration had resulted in loss of consciousness. This was entirely new ground.

When flight tests with the He 219 began, the objective was to safely fire a test subject from the aircraft at about 600 km/h. Eisenmann's He 219, which was coded DV + DI, had been equipped with an ejection seat for the trials. An experimental version of the seat, fitted with a dummy approximating the weight of a pilot, was installed behind the pilot in the place normally occupied by the observer's seat and was to be fired at a slight forward angle. The rear section of the canopy was removed for the experiments.

Eisenmann began the trials at a speed setting of 350 km/h. Each shot was filmed by two cameras mounted on the wings of the He 219 and by an escorting Do 217. A ground station also photographed each test shot. For this reason the aircraft could not be flying higher than 1,250 m when the seat was fired.

Initial difficulties centered around variations in the seat's flight path, which was in no way a function of weight and speed. The key point of this system, which operated on compressed air, was a delay valve which was responsible for the vital interval between the aircraft and the fired ejection seat, as well as the delay in opening the parachute. This escape system used

Erprobungsstelle Rechlin.
Oblt. Eisermann climbs into
He 219 W.Nr. 190113. This
machine was used for
various ejection seat tests,
resulting in the system being
declared operational.

a stabilizing chute, and once the pilot had released his harness and separated from the seat he activated his own parachute.

But the unpredictable delay seemed to have another cause. Sometimes the seat flew back over the empennage as it was supposed to, other times it struck the fuselage. During one test flight the dummy struck the tail section and nearly tore it half away. The result was an emergency landing. Not all of the seat could be found, which complicated the investigation. The top ends of the guide rails were often found to have broken off and were repeatedly strengthened, although it was Eisenmann's opinion that there was another reason.

The test flight that finally solved the puzzle almost cost the test pilot his life. That day Eisenmann approached the test field at 600 km/h and a height of 1,500 meters instead of the usual 1,250 meters, which probably saved his life. After the seat was fired the pilot saw nothing but white in his rear-view mirror. At the same moment he realized that the parachute was deploying and the seat was stuck on the guide rails. The He 219 pitched nose down and, with the parachute deployed, rapidly lost speed. Eisenmann was about to activate his own ejection seat when he realized that the presence of the rear seat made this impossible.

Then it suddenly came to him that a parachute would collapse if airflow was directed against it from the side. Eisenmann put the machine into a sideslip, applied full power to one engine and throttled the other back, and the white wall behind him collapsed. Later, while watching the film shot by the escorting machine, Eisenmann watched as his machine gradually returned from its stalled condition into a controlled dive and then, at approximately 100 meters above the ground, into horizontal flight.

He landed with the crumpled parachute on the tail section. Eisenmann finally had his proof: it was found that an error had been made in designing the system: a normal rectangular piston had been used instead of a slightly curved one. When the rails bowed, which always happened when the seat was suddenly struck by the slipstream just prior to leaving the rails, the piston immediately jammed in the cylinder.

The curved pistons were installed, after which everything proceeded smoothly. After numerous test shots with dummies, on 19 May 1944 the *E-Stelle Rechlin*'s test parachutist Buss was ejected from the aircraft. This and subsequent live tests by Buss were carried out over Müritz Lake, a large body of water. High-speed motor boats fished him out of the water. Buss pre-

Ejection seat system ready
for installation.

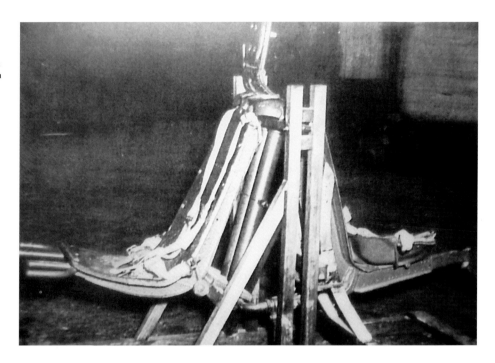

Ejection seats installed in an
He 219. Note the headrest on
the radar operator's seat at
the top of the photograph.
On later versions this was
attached to the canopy,
easing entry into the
cockpit.

ferred the soft landing in water to coming
down on land, having broken bones almost
twenty times during previous parachute
tests.

The remarkable first ejection from a
flying He 219 aroused much interest, however, an aircraft fitter had beaten test parachutist Buss to it. He inadvertently fired the
seat while the aircraft was still in the hangar. Luckily for him the pressure had
dropped in the ejection seat's compressed
air cylinder and was insufficient to carry
him to the hangar roof. He fell onto the tail
of the aircraft and suffered no serious injuries. As a result of this incident, on 19 May
the ejection seat's compressed air cylinder
was checked shortly before takeoff to prevent the shot from failing because of insufficient pressure.

Report:

Test parachutist Wilhelm Buss, 19/05/
1944. aircraft: He 219, speed 250 km/h,
jump height 1,000 m, ejection pressure 73
atmospheres. I found it a very tight fit when
I boarded the He 219, for the aircraft was
also a flying test-bed for various pieces of
test equipment and also carried all of the

equipment for night fighting. There was only centimeters of clearance between the equipment and my body. I strapped myself tightly into the seat. Someone checked me over and wished me good luck. The aircraft began to taxi. During the interval prior to takeoff I went over the procedure once more in my head. With so many safety precautions nothing could go wrong. I had installed time releases on my harness and on the parachute bag that would activate automatically in the event I struck the aircraft and was rendered unconscious. I had set the time release on the parachute bag to 15 seconds, the one on the seat harness to 10 seconds. In fact, however, after ejection I operated the seat harness and the parachute by hand, in each case in half the prescribed time, in order to show my comrades on the ground that I was uninjured.

We made a test pass to ensure that all the personnel were in position. The signal from below, a smoke signal from the boat, said: all set! During the second pass I fired a shot from the flare pistol and three seconds later left the aircraft. When ejected I was sitting with my back to the direction of flight. My feet were on the foot rests, and my hands grasped the handrails on the sides of the seat. When the ejection seat fired, my right elbow struck the cockpit sill. The impact was not painful. All that I can remember is something striking my right side. There was a loud bang when the seat fired, and steam and oil vapor from the compressed air system whirled up. There was a sudden jolt, which seemed gentler than in the preceding ground tests. In no time I was out of the aircraft and turning over. I found my entry into the air stream surprisingly gentle. It was the same feeling as in a free fall when the body has reached its terminal velocity and is being borne by the air. After leaving the aircraft I saw the fuselage of the He 219 pass beneath me. The tumbling movements were not uncomfortable. I could not determine the number

of my revolutions. The film later revealed four revolutions to a point just aft of the tail section.

I was satisfied to see how rigidly the body was held against the seat. After ejection my feet slid off the foot rests, perhaps because I had moved while tumbling, however, I put them back into position after the braking chute deployed. This can also be seen in the film taken by the escorting aircraft. However, the shape of the foot rests and my heel-less sports shoes helped my feet slip quickly from the foot rests. After the braking chute deployed I heard metal parts striking the seat's back rest and was then pulled gently into the suspended position. The noise was caused by the expansion of the spring which forced the braking chute out of its bag and clear of the dead air behind the seat, and also by the bearer rings of the braking chute on the seat mount. Remaining in the seat, I allowed myself to drop for several seconds. My view was very good. I found the sudden quiet to be pleasant. In the aircraft I had been sitting level with the two propellers, which had had an almost painful effect on my ears. I cast a quick glance at the open ribbon braking parachute; it had deployed perfectly. Measurements revealed that the seat descended at 39 meters per second while attached to the parachute. I now released the seat harness. I immediately fell from the seat. I let myself fall for four seconds and then pulled the ripcord for the main parachute. Its deployment was gentle. I looked for the seat and saw that it was drifting beneath my parachute. I followed it until it struck the water. The water immediately turned green, as I had attached a dye bag to the seat to ease recovery.

Shortly before landing in the water, I released the safety catch of the harness release box, and when I touched the surface of the water I activated the release by hand pressure. The parachute, still being blown by the wind, pulled the harness from my body and landed several meters away. As I had landed among a group of rocks, I tried to swim to land. At the first strokes I felt pain and tension in my right wrist and saw that I was bleeding. Finally I was pulled out of the water into an inflatable boat. The hand grips were subsequently moved forward so that the arms could not flex when the seat was fired.

Third ejection:

The third ejection was involuntary: the camera aircraft was flying on the right side. I intended to give the usual signal by firing a flare, with ejection following three seconds later. The cloud layer was at 1,000 meters and was only partly broken. Visibility was thus restricted.

I was sitting with my head over the cockpit sill, trying to acquire visual contact with the ground. I was holding the pistol in my right hand, safety off, my finger not on the trigger, when the pilot pulled the release lever and I unexpectedly found myself outside the aircraft. There was a jolt under my rear, after which the seat tumbled and spun. As I had been looking down over my right arm and the cockpit sill, my head struck my right arm and the arm with the pistol hit my thigh. The braking chute automatically deployed as I was regaining my senses. I was still holding the flare pistol in my right hand. As it was bothering me, I threw it away.

I let myself fall in the seat, which was still attached to the braking chute, for some time, and finally operated the release by hand. The seat fell away immediately, after which I opened the main parachute. I came down on the airfield grounds. The film showed that my arms were outstretched when the seat fired and that the seat did not revolve on a single plane. Instead, because of my body's extended position and outstretched arms, it whirled through the air,

turning in every direction. Use of a back belt would have proved useful here. Experience has shown that the harness attachment point on the back of the seat should be at shoulder height, which would prevent the upper body from bending forward.

This jump, on 13 June 1944, was the last one necessary to fulfill acceptance requirements. The ejection seat could now go into series production.

Testing of the He 219 continued, and prototypes V7, V8, and V9 were flown to Venlo. There, in the front-line maintenance facility, the He 219 A-0's initial armament package was finalized. On 22 May 1942 test pilot Schäfer carried out live firing over the Zuider Zee with the He 219 V7 and V9. During the V9's flight the aircraft suffered a serious loss of coolant, as a result of which the port engine had to be shut down. Schäfer flew back to Venlo on one engine, restarted the shut-down motor and landed safely.

Armed with two MG 151/20s in the wing roots and four MK 108s in the ventral tray, the prototypes received the designation He 219 A-0/R1. Later the MK 108s were replaced with long-barreled MK 103s, which increased maximum attack range to 1,200 meters. This version was designated the He 219 A-0/R2 and was the first to be fitted with the FuG 212.

The radar installation presented some difficulties at the beginning, since there was insufficient space in the nose in front of the cockpit. Heinkel did have another A-0 with an enlarged nose under construction at Schwechat, however, in cooperation with Telefunken the Lichtenstein set was modified to allow it to be installed in the standard He 219. At Venlo a comparison flight was made with a Bf 110 G, the results of which are documented below:

Report by Tech.Ing. Maciejewski Currently Stationed at Venlo

Report He 219 V7, V8, V9
18/5/1943

Aircraft V9 was ferried to Venlo by Herr Schäfer on 15 May 1943.

On 15 May a comparison flight was carried out between the V8 and an Me 110 G. The V8 was approximately 10 km/h faster. The Me 110 G was a trimmed machine, meaning it was flown with no camouflage paint or armament. A calibration flight with the V7 revealed a speed of 440 km/h at ground level at 1.3 atmospheres of boost and 2,300 rpm. Flame dampers were installed on the V7 during the night of the 15th-16th. The installation resulted in a speed loss of 18 km/h.

On all machines it was observed that the forward edges of the split flaps bent by up to 20 mm along their entire length at high speed. Director Francke was aware of this, and he promised new strengthened flaps. These were delivered by air by Herr Meschkat on 17 May. The new flaps were installed on the evening of 17 May. The V7 fired its guns in the air for the first time on the 17th, resulting in the following unusual manifestation. On the left side of the fuselage, near the observer's position and abeam the wing-mounted weapons, 30 to 40 scratches appeared in several places,

causing paint to flake away and the skinning to bow inward. It was first suspected that shells had grazed the fuselage, however, tests in the firing butts revealed that there was sufficient clearance. It was thought that when the shells' guide rings disintegrated they were thrown against the side of the fuselage, perhaps under the influence of the propeller. As well, the entry ladder dropped and was slightly damaged near its lower end by a shell fired by the fuselage weapons. The shell deflector prevented further damage. The bore sight correction mechanism revealed that there was 50 mm clearance from the entry ladder. Thought was given to whether the shell deflector should be set even shorter. Deployment of the entry ladder was attributed to the locking pin, which on this machine was rather too short.

The high rate of attrition of tachometers was also a cause for concern. Oblt. Hausdorf, GeTO, proposed that the ammunition canisters be armored against fire from ahead and ahead and below. Proposed was a vertical armor plate in front of the canisters and an armor plate in the ammunition canister cover plate.

Maj. Streib also requested an emergency fuel jettisoning system to allow the pilot to retain control of a fully-loaded machine after the loss of one engine. Special-

Artist's impression of an He 219 A-0 as it entered service with NJG 1 in the spring of 1943. It still carries the antenna array, or "Mattress" associated with the FuG 202 Lichtenstein BC radar, whose effectiveness was severely hampered by Bomber Command's use of Window beginning in July 1943.

ists from the Patin and LGW companies checked and reset the autopilots of the V7 and V9.

The Lichtenstein system had to be completely reset on all three machines. In the process, numerous incorrect settings were discovered. The basic propeller setting on the V9 was slightly off, with 2,100 rpm instead of 2,500 at 1.3 atmospheres and 12.10 minutes.

No significant complaints were raised during in-flight trials. Maj. Streib found the flame damper installation to be completely satisfactory.

Placement of the shell ejection chutes on the bottom of the fuselage resulted in damage to the antenna for the FuG 16, and relocation of the antenna mast was a definite consideration. Air firing trials were continued on 18 May.

On 30 May *General* Kammhuber and *General* Peltz (Commander of the Air War against England) arrived at NJG 1's base at Venlo to inspect the Heinkel machines. The V7 was examined closely and received very favorable comment. The He 219's endurance, in particular, was praised, this ranging from four to five hours depending on power settings. *General* Kammhuber decided that the V9 would be the first to fly a live sortie, which was planned for the week of 2 to 5 June. The V7 was to be used for training at first. At this meeting *General* Kammhuber asked for an He 219 to use as his personal aircraft. Heinkel subsequently gave consideration to converting a prototype for the personal transport role.

In a development summary dated 10 June 1943 Heinkel designers reported: "The importance of an oblique weapons installation for the night fighter role was stressed. What they have in mind is a rigidly-mounted armament set installed at a forward angle of 60 to 70 degrees." It went on to say: "Use of GM 1 injection, enlarged wings, and the power plants DB 603 G, DB 603 G with injection, DB 603 G with T11, Jumo 222, DB 627, and BMW 803 S is being investigated at Rechlin and by Daimler Benz."

Operations and Further Trials

On 11 June 1943, one week later than planned, the He 219 V9, bearing the code G9+FB, was readied for the first live night fighter sortie. That night 783 bombers took off from bases in England to attack Düsseldorf. The attack force included four-engined Halifax and Stirling bombers, the latest Lancasters, and obsolescent Wellingtons. 693 bombers found the target and dropped their bombs, guided by de Havilland Mosquito pathfinders equipped with Oboe target-finding radar. The raid left 120,000 residents of Düsseldorf homeless, and many were killed or injured.

Every serviceable German night fighter was sent aloft to intercept the raiders. The He 219 A-0/R2 crewed by *Major* Werner Streib and Uffz. Helmut Fischer flew over the Netherlands. Guided to the bomber stream by ground controllers, Fischer then used his Lichtenstein BC to position his pilot behind the bombers. Again and again Fischer guided Streib into favorable attack positions, and five heavy bombers fell to the Heinkel's cannon.

In spite of the strong German response, the old city of Düsseldorf was burnt almost completely to the ground. No fewer than 38 RAF bombers failed to return that night, and the crew of Streib and Fischer accounted for 1/8 of these.

Streib radioed that his ammunition was gone. He approached home base with fuel running low and several instruments unserviceable. While on approach to land the canopy misted over, as often happened,

Photographs taken from the accident report on the He 219 V9 (G9 + FB), which was wrecked on landing after its sensational operational debut in the hands of Maj. Streib.

and Streib was forced to go on instruments. Streib lowered the electrically-operated flaps to the landing position and then lowered the undercarriage. Then, unnoticed by the pilot, the flaps, which had not locked down, returned to the normal position.

The He 219 dropped like a stone and struck the runway at Venlo with great force while traveling at excessive speed. The shock of landing was so great that the tires blew, the force was transmitted to the airframe, and the starboard engine was torn from its mounts. The aircraft literally disintegrated. The entire cockpit broke off at the wing roots and slid more than 50 meters along the concrete runway. None of the eyewitnesses imagined that the crew could survive, however, Streib and Fischer escaped almost uninjured.

On 12 June 1943 the following message, addressed to Prof. Dr. Heinkel, was received by the Rostock teletype office:

Your He 219 saw its first live action during the night of 11-12 June 1943. On this first mission with the He 219 Major Streib scored five night victories. The He 219 has thus proved its special qualities as a night fighter against the enemy. My thanks and appreciation to you, your designers, and workers. Unfortunately, the machine crashed on landing. I ask that you follow up with all means to see that further deliveries of the He 219 are accelerated.

Signed Kammhuber
*General der Flieger
and commanding general of the
XII. Fliegerkorps
and General der Nachtjagdflieger.*

This success obviously had little effect on *Generalfeldmarschall* Milch. On 15 June, during a conference with the GL/C he tried to downplay the success of the He 219. Among his remarks: "The He 219 is good, it shot down five in one mission. I can't ask for more than that. But perhaps Streib would have had just as much success with another machine."

On 1 July *Hauptmann* Frank took over command of I./NJG 1 and began flying an He 219 A-0. He was able to continue the run of success begun by Streib, now *Geschwaderkommodore*. On 26 July he shot down two British bombers, followed by one on the 27th, one on 23 August, and three more on the 31st. Frank could have accounted for even more enemy aircraft on 27 July, however, on four occasions his cannon failed to fire, and in each case he was forced to abandon the pursuit.

Testing of the He 219 A-0/R3 (GE + FN) was meanwhile proceeding at a rapid pace. It was realized that there would be difficulties with the belly armament. Deliveries of the MK 103 by Rheinmetall-Borsig were slow, and the manufacturer was forced to turn to the short-barreled MK 108 for some production aircraft. Later, production difficulties were also encountered with the MK 108, and in many cases aircraft were armed with MG 151/20 cannon in the ventral tray.

At about this time a new camouflage scheme for night fighters was tested for the first time. The former washable black night finish was replaced by an overall pale gray scheme (RLM 76). Although it is generally believed that German night fighters were

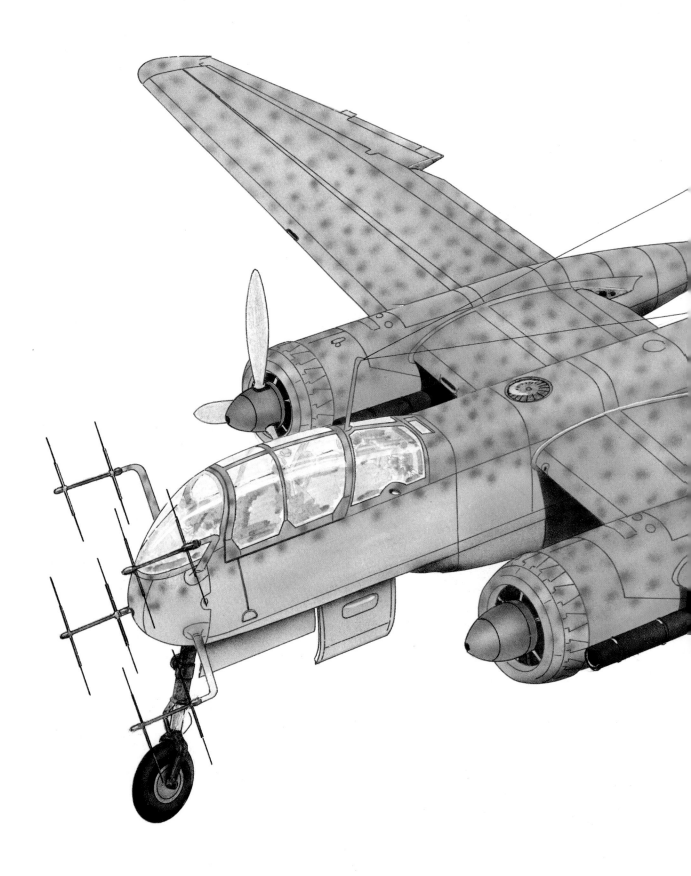

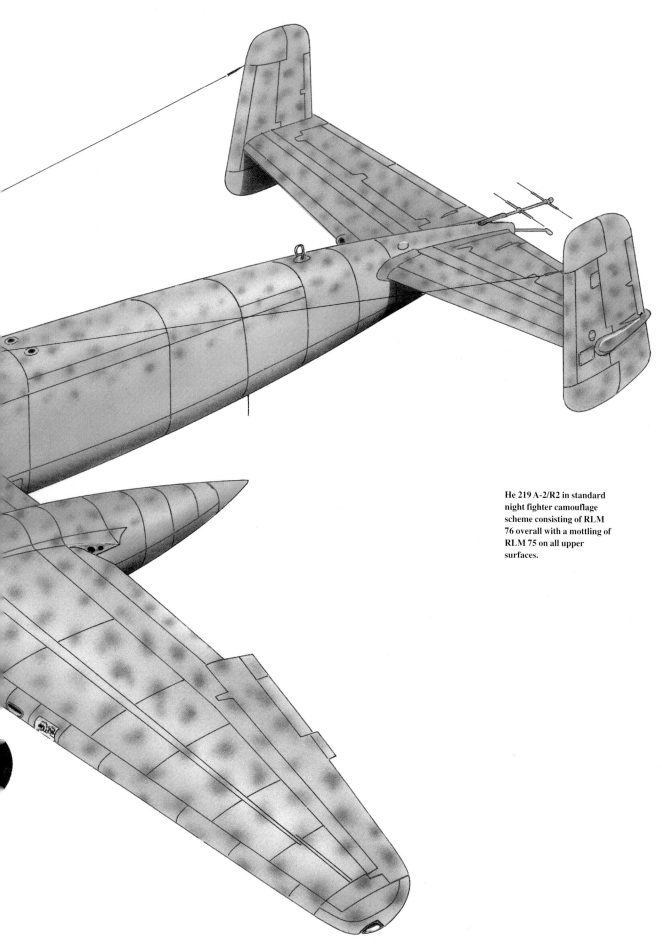

He 219 A-2/R2 in standard
night fighter camouflage
scheme consisting of RLM
76 overall with a mottling of
RLM 75 on all upper
surfaces.

finished exclusively in these colors, they in fact operated in a wide variety of camouflage schemes. In September 1943 the technical services at Venlo made the Heinkel Werk aware of the new style of *Balkenkreuz* to be worn on the fuselage sides and upper wing surfaces. From then on all He 219s were delivered with white outline crosses. From the middle of 1943 in particular, the imagination of the painters in the front-line distribution centers and unit workshops knew no bounds. Even within the *Staffeln* the He 219 was finished in a wide variety of camouflage schemes. Most sported finishes consisting of spots, patches, and irregular areas in RLM 02, 74, and 75 sprayed over RLM 76, or in some cases RLM 02. Overspraying in the darker colors had the effect of darkening the base color. From spring to autumn 1944 there were cases of aircraft which had their starboard wing and engine nacelle, with the exception of the leading edge and wingtip, painted matte black. This was intended as an identification feature for German anti-aircraft guns and searchlights. From the end of 1944 it was common for aircraft engaged in night close-support missions to have their wing and fuselage undersides painted black.

As a rule, the codes worn by He 219 night fighters followed the pattern of other combat units. Since the majority of *Uhus* served with the *Stab* and *I. Gruppe* of NJG 1, with only a few examples delivered to the *NJ-Staffel Finnland und Norwegen*, II./NJG 1, Erg./NJG 2 and NJGr. 10, the list of unit codes is a brief one. These consisted of letter-number combinations applied forward of the fuselage *Balkenkreuz* with characters roughly one-quarter the height of the cross, usually in black.

B4 + *NJ-Staffel Finnland und Norwegen* (later *NJ-Staffel Norwegen*)
G9 + NJG 1
R4 + Erg./NJG 2 (later also 4R +)

I./NJGr. 10 used the types of code employed by day fighter units, however, to date there is no documentary of photographic proof that the He 219 and Ta 152 also wore these. Nor are there any photographs or documents to confirm the codes worn by the machines of Erg./NJG 2 and the *NJ-Staffel Finnland und Norwegen*.

Aft of the fuselage Balkenkreuz was a combination of two letters. The first letter indicated the aircraft's sequence within the *Staffel*, with the A usually reserved for the *Kommodore*, *Kommandeur*, or *Staffelkapitän*. To prevent confusion, the letters G, I, J, O, and Q were not supposed to be used, though here too "exceptions were the rule." In I./NJG 1's handwritten loss list an He 219 A-0 coded G9 + OK is recorded under 9 September 1944. These letters were either painted in the *Staffel* color or black outlined in the *Staffel* color.

A *Geschwaderstab* (green letters)
B *Stab I. Gruppe* (blue letters)
C *Stab II. Gruppe* (blue letters)

The last letter indicated the *Staffel* number within the *Geschwader*. The three *Staffeln* within the *Gruppen* wore the iden-

I. Gruppe	II. Gruppe	III. Gruppe
H *1. Staffel*	M *4. Staffel*	R *7. Staffel*
K *2. Staffel*	N *5. Staffel*	S *8. Staffel*
L *3. Staffel*	P *6. Staffel*	T *9. Staffel*

tification colors white (*1., 4.* and *7. Staffel*), red (*2., 5.* and *8. Staffel*) or yellow (*3., 6.* and *9. Staffel*).

Combinations of four letters appeared only on various prototypes used by Heinkel in its test program, or as call signs within the service trials units and the *E-Stellen* at Rechlin and Augsburg.

In July 1943 Ernst Heinkel Flugzeugwerke conducted various experiments aimed at increasing the range of the He 219. The RLM considered using the aircraft against England armed with one 500-kg bomb. Calculations revealed that the use of extra internal tanks would result in an increase in range from 2,400 km to 3,350 km. A range of 3,750 km would be possible with external drop tanks.

As called for in the test planning, the crew of Könitzer and Consten carried out various diving flights with the He 219 V2 in order to determine the type's limiting maximum speed. On 10 July 1943 the machine crashed and the first life was lost in an He 219. Engineer Consten, who flew as observer, reported: "Our mission was to carry out two diving flights at 675 km/h and 700 km/h, switching on the tail unbalance at an altitude of between 4,000 and 3,500 meters. The climb to 6,400 was completely normal. Unlike previous flights, when Mr. Könitzer eased the aircraft into an approximately 15° dive, this time it began abruptly. The dive was too steep and felt like 30°. The speed of the first calibration flight was reached at a height of 4,000 m. I called to the pilot to reduce speed. His voice was choked, and I believe he said something to the effect of, 'I can't reduce speed.' Communications inside the machine were extremely poor. At about 3,500 m the speed was 700 km/h and the airspeed indicator reached the stop (750 km/h). I had the impression that the dive angle remained constant. Approximately 10 sec-onds after passing 675 km/h there was a bang, and a jolt shook the machine (the tail section broke off, author's note). I lost consciousness. When I came to, I had the feeling that the machine was in a vertical dive. I was twisting and turning and trying to jettison the canopy when the machine rolled onto its back. I reached for the eject handle, but I wasn't sure that I had jettisoned the canopy. Then I lost consciousness again. When I came to I was lying on the ground." The accident investigation found that the pilot had been unable to recover from the dive due to the great elevator forces and had deployed the braking chute. As a result of the abrupt acceleration of the pilot along the aircraft's longitudinal axis, the control column was pushed forward, initiating an outside loop that caused the crew to lose consciousness. The braking chute and the tail section must have been torn off soon afterwards. The aircraft crashed approximately 1.4 kilometers southwest of Mühl-leiten in the Loban Forest. On impact the pilot was thrown from the machine with his ejection seat and was killed.

There now followed a perplexing series of designations for the He 219 series aircraft. The A-1 series, which was planned as a high-speed bomber, was canceled before construction began. The first fighter version to be delivered was in fact identical to the A-2, however, it was (because it came from the pre-production series) designated the A-0. The units were issued pilot's notes and operating manuals for the He 219 A-0 and, from November 1944, the A-7 (see the survey of variants in the appendices).

Quantity production had finally begun at Rostock on 3 August 1943. This measure was taken on account of production delays

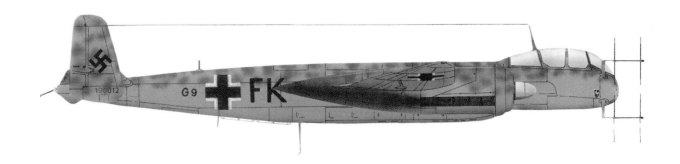

Heinkel He 219 A-0/R6

Operational aircraft of 2./NJG 1 based at Venlo in March 1944 and flown by Oblt. Ernst-Wilhelm Modrow.

While flying the He 219, in just three months Modrow shot down 22 enemy aircraft at night, including one Mosquito.

Aircraft were delivered with the flame-damper tubes painted black, however, the paint burned off soon after operations began and the tubes rusted very quickly.

The DB 603 A engines usually produced heavy exhaust staining along the engine nacelles back to the tailplanes. The same phenomenon can be seen on Do 217 N aircraft. Exhaust staining also affected the main undercarriage doors.

Geschwader emblem worn by all night fighter units.

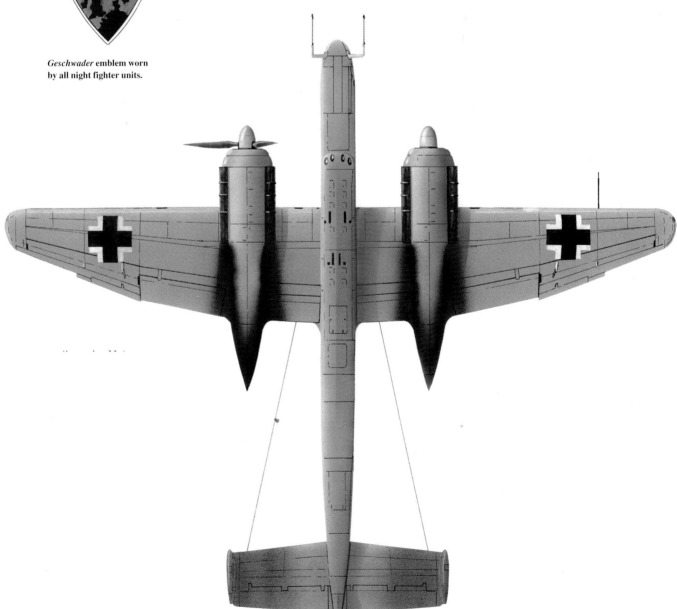

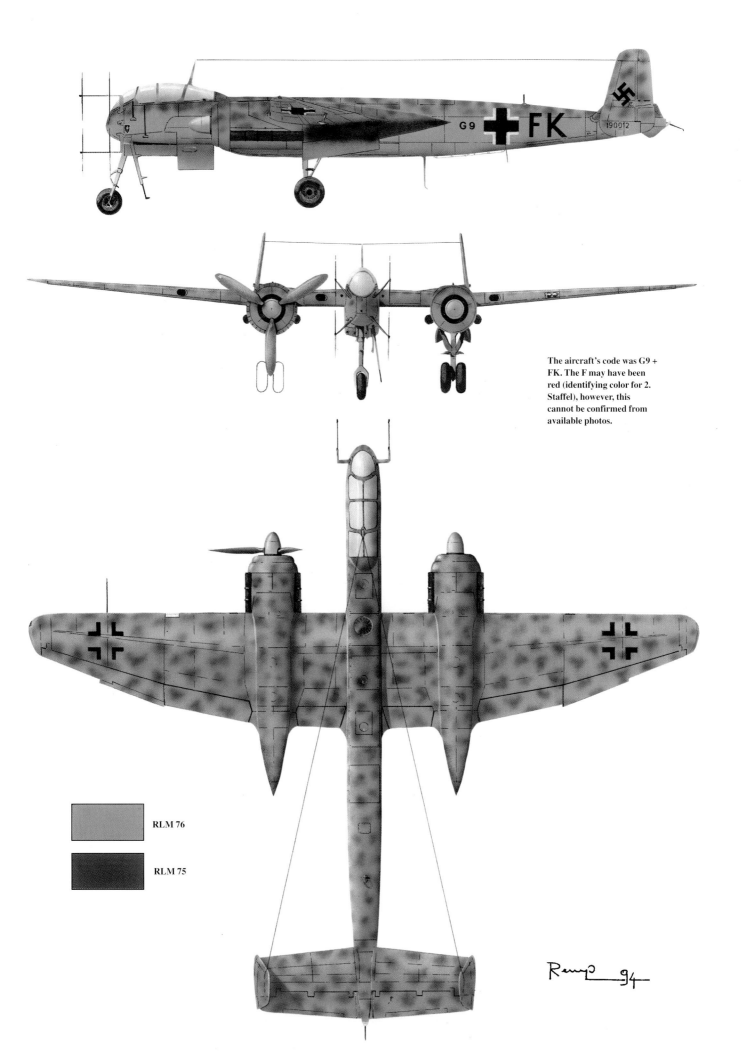

The aircraft's code was G9 +
FK. The F may have been
red (identifying color for 2.
Staffel), however, this
cannot be confirmed from
available photos.

RLM 76

RLM 75

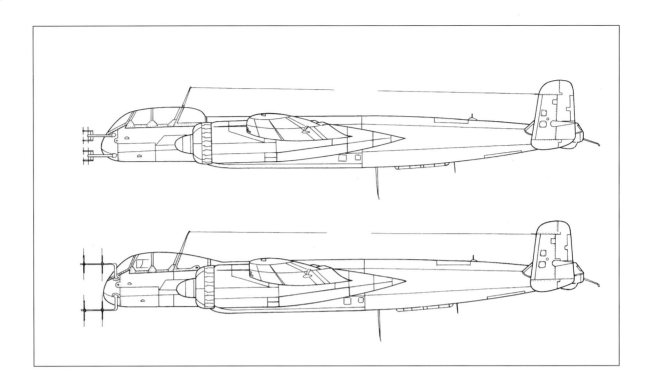

He 219 A-09, a pre-production aircraft equipped with the FuG 212. It was with this type that Maj. Streib conducted service trials.

He 219 A-0/A-2 production aircraft equipped with the FuG 220. The two models were identical, and the designation changed only with the number of aircraft built.

at Mielec and Budzyn. At Rostock the He 219 was produced alongside the He 111 H-20. RLM Program 223 of 15 April 1943 called for a monthly output of fifty He 219s. At this point the navy also began showing interest in the type. This inquiry contributed to Heinkel's hopes that he might be allowed to construct as many as 300 machines per month, as he stated on 29 June. The He 219 V16, W*erknummer* 190016, was completed on 10 August 1943. This was the first *Uhu* to carry the improved FuG 220 Lichtenstein SN-2 airborne radar. Telefunken wished to carry out future trials with the A-0 production machine at its company airfield at Berlin-Diepensee. On instructions from the RLM, work on the He 219 A-0 *Werknummer* 190008 was accelerated to allow installation of the airborne radar to be completed. The aircraft was supposed to have an increased range, and for this reason the company awaited eagerly for the two armored and enlarged tapered tanks which were supposed to be installed as engine nacelles.

At the state secretary's conference on 20 August 1943, Milch referred to the Ju

188 as being at least as good a night fighter as the He 219. Consideration was being given to converting some production space at Henschel, which built the Me 410, in order to produce the Ju 188 there. Since a shortage of BMW 801 power plants was expected, the Ju 188 was to be converted to accept the DB 603. Alarm bells sounded at Heinkel, for such a move would hamper the delivery of DB 603 engines for its He 219 even more than was already the case. Heinkel director Francke tried to explain the importance of the power plant question for the He 219 in a memorandum. According to a statement by Streib, the He 219 had become too heavy and clumsy, and Dir. Francke, not least because of Streib's comments, foresaw that a switch to the DB 627 or DB 628 would be absolutely necessary.

On 26 August 1943 a conference was held in Schwechat, at which the directives for a high-altitude variant of the *Uhu*—the He 219 B-1—were laid down. This was supposed to be equipped with a wooden wing and powered by 2,500 hp Jumo 222 A/F or E/F engines, which would bestow a maximum speed of 615 km/h at sea level

74

and 759 km/h at 12,500 meters. A range of 2,850 kilometers appeared possible with a fuel consumption of 3,400 liters.

On 27 August, after its FuG 220 had been calibrated, the He 219 A-01 was flown to Diepensee. The V19 was completed in the same month. It was an aircraft of the A-0 pre-production series with *Rüstsatz R6*, the designation for two upwards-firing MK 108 cannon. The aircraft's ventral armament was reduced to two MK 108s. Under way at the same time was a highly-secret project: an He 219 powered by two DB 603 Gs and a BMW 003 turbojet engine producing 650 kg of thrust. It resulted in the following performance estimates:

Range: 1,700 km
Speed for 30 minutes: 700 km/h

The He 219 V10 and V12 saw action with I./NJG 1 on the night of 30-31 August 1943. Haup*tmann* Frank, who flew the V12, and *Oberleutnant* Strüning in the V10 each scored three victories. The V12 was hit in the fuselage and engine during its second attack, forcing Frank to shut down one engine. In spite of this he attacked and shot down a third enemy bomber.

As a result of the success of the Heinkel night fighter, I./NJG 1 pushed for the delivery of more machines. The V3 was ordered to Venlo instead of going to Tarnewitz for weapons trials. A telex message to the RLM dated 2 September 1943 listed the aircraft which were assigned to Venlo by 12 August:

A-01 Located in Diepensee for the purpose of calibrating the FuG 220 system by Telefunken. To be completed today and subsequently ferried to Venlo.

A-03 FuG 220 installation completed. Ready for delivery for two days, however bad weather made this impossible. To be ferried today.

A-04 FuG 220 installation completed, test flight tonight after which it will be ferried to Diepensee.

A-05 FuG 220 completed, aircraft flying, will probably be ferried to Diepensee on Sunday, 5 September 1943.

A-06 FuG 220 completed, aircraft flying, will probably be ferried to Diepensee on Sunday, 6 September 1943.

A-07 FuG 220 to be completed on Friday, 3 September 1943, after

The He 219 A-016 with the prototype installation of the FuG 220 Lichtenstein SN-2.

The He 219 V33 was used by Telefunken for equipment and antenna testing. This photograph is believed to have been taken at Munich-Riem.

which it will begin flying. Firing, handover, test flight. To be ferried to Diepensee on Wednesday, 8 September 1943.

On 2 September 1943 Ernst Heinkel Flugzeugwerke released an overview of planned deliveries of the He 219. A total of 120 prototypes were to be built by the end of July 1944.

During the night of 5-6 September 1943 the He 219 V10, which was stationed at Venlo, was lost in a crash. The fighter attacked a British bomber, but its first burst missed the target. Alerted to the threat, the bomber's gunners opened fire and the Heinkel received hits, including some from below. A bullet had severed the control cable from the fuel tank selector lever, and during the flight back to base the pilot was unable to switch to Tank 1. The aircraft's engines quit one after the other. The crew of Strüning and Bleier decided to abandon the machine. The canopy jettisoned without difficulty, however, the ejection seats would not fire. Strüning pulled the aircraft

up to enable his radio/radar operator to get out, and then climbed out himself. He struck the antenna mast and tail surfaces, sustaining bruised ribs and contusions. Radio/radar operator Obfw. Bleier was initially reported missing, but was later found dead.

On 22 September 1943 the flight program for the turbojet-boosted He 219 A-010 was finalized. Prof. Ernst Heinkel had promised the workers 180 bottles of wine if they completed the aircraft ahead of schedule. By the end of September the He 219 A-010/TL reached 539 km/h at low level. The installation of flame dampers resulted in a 9 kilometer-per-hour loss of airspeed.

Hanover was the target for 678 RAF bombers on the night of 27-28 September 1943. Of these 612 dropped their bombs, but not over the city. The British crews were not familiar enough with the new H2S navigational radar, and features of the local terrain caused them problems. The Steinhuder See, a large lake several kilometers square, which the British used as a way point, had been almost completely covered with boards and nets. This saved the city of Hanover—at least on this night. The bombs fell in open country, and only a few landed in small villages. Millions of strips of aluminum foil, Window to the British, were dropped in an effort to blind the German defenses. The *Freya*, *Würzburg-Riese*, and *Lichtenstein BC* radars were blinded. Nevertheless, 38 bombers failed to return. That night Hans-Dieter Frank, the *Kommandeur* of I./NJG 1, took off to intercept the enemy bombers in He 219 A-03, *Werknummer* 190055, G9 + CB. He had scored another victory on the sixth of the month. Approximately 25 kilometers northwest of Celle his He 219 collided with a Bf 110 G-4 of the *Geschwaderstab* of NJG 1. As they went down, the two machines separated and crashed five kilometers south of Bergen, near Meisendorf. *Hauptmann* Frank, who three months earlier had received the Knight's Cross while *Staffelkapitän* of 2./ NJG 1 and who had 56 victories to his credit, ejected from the aircraft. Frank landed smoothly, however, he was choked to death by the cable of his radio-helmet, which he had forgotten to disconnect and which compressed and crushed his larynx. His radio/radar operator Obfw. Erich Gotter was either thrown from the machine or bailed out without the aid of his ejection seat, for the seat was found in the wreck with the harness undone. Gotter's body was found later. The Bf 110 (G9 + DA) crashed two kilometers from the He 219; all aboard, Hptm. Friedrich, Oblt. Gerber, and Obgefr. Weisske, were killed.

On 24 September, during a mock combat with a Bf 110, Hptm. Förster was involved in a near fatal incident flying the He 219 V7. While in a 50° to 60° turn at 260 km/h at an altitude of 2,200 meters, the aircraft suddenly dropped its port wing and went into a steep spiral. The machine had shuddered slightly about 360° prior to the stall. The control column was forced against Förster's body, and it required all his strength to push it forward. He cut the throttles, centralized aileron and rudder, and after four revolutions he was able to bring the machine out of its dive. The aircraft returned to a normal attitude at a height of approximately 900 meters. *Hauptmann* Förster later stated that he never considered abandoning the aircraft.

By 10 October 1943 the technical field service of the Ernst Heinkel Flugzeugwerke reported various problems with the He 219 night fighters, one of which was alarming: fully-fueled aircraft were observed to be streaming fuel vapor from the vents on the fuselage sides shortly after takeoff. The fuel ran along the fuselage and entered every opening, especially the aft footstep, pooling between the fuselage bulkheads. The mixture of fuel and electrical equipment in the fuselage were a potentially explosive combination. It was not known if any aircraft had been lost under these circumstances, however, immediate steps were taken to alleviate the problem.

On 20 October 1943 *Stab*/NJG 1 lost its He 219 A-04. The weather was rainy, and there was the threat of icing. At 2130 hours Lt. Schön informed ground control

Oberleutnant **Modrow's He 219 A-0, G9 + FK, of 2./NJG 1 in formation flight with Oblt. Baake, whose radar operator, Uffz. Waldbauer, took this photograph.**

that he had contact with enemy four-engined bombers and then suddenly disappeared. On the morning of 21 October residents of the town of Storbeck, near Stendal, found G9 + CB, *Werknummer* 190054. The bodies of Lt. Schön and Uffz. Marzotke were found two kilometers from the crash site near Gross Ballerstedt. Both were lying near their ejection seats. The machine had crashed into a field at high speed in a nearly vertical attitude, creating a crater three meters deep and eight meters across.

Hauptmann Meurer, a wearer of the Knight's Cross with Oak Leaves, now assumed command of *I. Gruppe* of NJG 1. He was considered a very practical-minded officer, and he was the He 219's and Heinkel's chief supporter. By 13 October there were seven He 219s at Venlo, however, only two were serviceable. The men of I./NJG 1 were very much supporters of the He 219, mainly because it was the only available night fighter with the required performance. Kammhuber's tendency to request great endurance for the He 219 had proved correct, especially for the "*Wilde Sau*" role. The crews flew missions from Venlo over the Hanover—Rostock—Stettin area or over central Germany to Munich. They also placed great value on being able to return to their home base, because only there could their He 219s be made ready for action again in a short time. One crew flew back to Venlo after losing an engine over Munich.

On the night of 18 October the German air defenses claimed 17 enemy bombers during an attack on Hanover. One of these went to *Hauptmann* Meurer, who that night flew He 219 A-0 G9 + BB. He shot down another British bomber on 23 October, and one more during the night of 3-4 November 1943. The target that night was Düsseldorf, which was attacked by 527 heavy bombers. Meurer's victory was the last by an He 219 in 1943.

In October I./NJG 3, commanded by Hptm. Prince zu Lippe-Weißenfels, had sent a detachment of six ground personnel to Venlo for familiarization with the He 219. The unit expected to receive its own He 219 A-0s in the near future.

In the first weeks of November approximately fifty German night fighters were equipped with the improved SN-2 radar. This equipment was relatively immune to Window, however, only twelve machines and crews were fit for operations. The main reason for this was delays in training operators to use the complicated and sensitive equipment. During this period II./NJG 1 at Twenthe and the various *Staffeln* of I./NJG 2 at Gütersloh, Kassel, and Neuruppin enjoyed little success. The He 219 was unable to do little to improve this situation, however. I./NJG 1 still had just seven *Uhus*, and none of these were operational. The He 219s were subjected to a series of tests which revealed minor shortcomings, however, these were enough to keep the aircraft out of combat. Flights at high altitude revealed that the cockpit heating system was virtually useless. Frosted cockpit glazing made accurate shooting impossible, even from favorable firing positions. The radio/radar operator could guide his pilot to enemy aircraft, but when the latter looked up from his instruments to acquire visual contact all he saw was ice patterns, forcing him to fire blind. Since frequent misting over of the canopy made landings extremely dangerous, even for experienced pilots, *Hauptmann* Meurer considered banning younger pilots from flying

Retouched photograph depicting the A-010/TL with underslung jet engine.

the type until the problem was solved. For a time, the service test unit was unable to obtain spare parts. Tires were the only items available through the normal supply channels. With larger scale use of the He 219 approaching, this was one of the burning questions that had to be solved. Some machines were taken out of service and stripped of parts in order to keep others airworthy. On 5 November it was decided to ferry all of NJG 1's *Uhus* to Rechlin to once and for all solve all the problems that had arisen and restore the aircraft's combat readiness.

Testing of the He 219 A-010/TL went on, however, there was no improvement in the type's low-level maximum speed of 530 km/h. The aircraft proved to be extremely tail-heavy when the turbojet engine was operating, and dynamic pressure was very high, resulting in relatively high control forces.

On 13 November 1943 the He 219 A-010 sustained 40 percent damage in a crash-landing at Aspern. The crew had instructions to start the turbojet engine at various altitudes, and the second attempt resulted in a heavy flame buildup in the power plant. The crew thought that the entire engine was on fire and immediately shut off the engine and the fuel cock. Soon afterwards the piston engines also stopped, and the crew carried out an unpowered forced landing. By this time the aircraft had achieved a speed of 622 km/h at a height of 6,000 meters, which was about 70 km/h greater than the standard A-0 model. After repairs, the machine was supposed to resume the test program.

Thirty He 219 A-0s left the Schwechat factory by 1 December 1943. It is likely that this figure included the first twelve true prototypes. Planned production now totaled 1,093 aircraft. By this time Mielec had become purely a subcontractor facility. The pre-production series, which was to consist of 183 He 219 A-0s, was now under construction at Marienehe, where the 715 A-1s were also to be built. A further 165 He 219 A-0s were to be built in Vienna, and there was a possibility that this total might be increased to 195.

On 3 December State Secretary Milch made another attempt to halt production of the He 219. At Milch's urging the GL/C-E2 made three proposals concerning the night fighter program. These revealed Milch's desire to have Heinkel participate in the Junkers production program. The three proposals were:

1. Wind down series production of the He 219, switch Heinkel Group North to the Ju 88 G, Heinkel Group South to the Do 335.

2. Production of the He 219 to be reduced from 100 to 50 machines per month immediately, to be built by Heinkel Group South. Heinkel Group North to build 50 Ju 88 Gs per month.

3. No more than 100 He 219s to be built per month.

It should be noted that the Ju 88 G was only in the experimental stage at this point, and that there was not even a prototype of the Do 335 A-10, while quantity production of the He 219 was already under way!

On 8 December Ernst Heinkel Flugzeugwerke reported seven new He 219s ready for front-line service. After completing check flights, these seven aircraft (A-013, A-015, A-016, A-017, A-020, A-022, and A-025) were supposed to be issued to NJG 1. Bad weather caused delays. The check flights were ordered on account of recurring problems with the machines issued to the combat unit. After leaving the factory each machine had to complete a two-hour program, which consisted of the following:

1. Climb to 7,000 meters during which the pilot made a complete check of the engines, while the radio/radar operator checked the FuG 16 and FuG 10 radios, the low-frequency air-ground radio, and the DF set.

2. Two machines rendezvoused at a designated point to carry out a practice interception with the Lichtenstein SN-2.

3. Descent was combined with an approach using the automatic homing set.

4. The aircraft was leveled off at 2,000 meters, and the pilot checked the machine's handling characteristics (control forces, control effectiveness, etc.).

5. Shortly before landing the aircraft was flown over the VHF radio beacon (height approximately 200 m) in order to check the FuG 101 radio altimeter.

At a development conference held by the GL/C on 13 December 1943, the *General der Jagdflieger* demanded that the high-altitude variant of the He 219 be developed with GM 1 injection in addition to the DB 603 G supercharged engine. The

In April 1944 *Oberleutnant*
Modrow flew G9 + FK from
Venlo. In addition to the
FuG 220 Lichtenstein SN-2,
his He 219 A-0 carried the
FuG 212 C1 with centrally-
mounted antenna. Also
interesting is the black-
painted area on the underside of the starboard wing,
an identification marking for
German flak gunners.

The only known photo of an aircraft of Nachtjagdgruppe 10 at Lärz. It depicts mechanic Hermann Klauser standing on the nosewheel of an He 219 A-0. Noteworthy details include the night fighter emblem on the side of the nose and the letter F and number 05 beneath the FuG 212 antenna.

The Focke Wulf Ta 154 Moskito was planned as the German wooden equivalent of the British Mosquito. Unsuitable glues caused the bonds between wooden components to deteriorate quickly and prevented the type from entering service with front-line units. Depicted here is the third aircraft of the A-0 pre-production series, TQ + XE, W.Nr. 120005, which was tested by NJGr. 10 with SN-2 radar.

oblique armament was now to consist of a pair of MK 108 cannon. It remained to be determined whether and in what quantities the large master-slave compass demanded by NJG 1 would be available.

The RLM had obviously been planning for some time to supplement the designations of German combat aircraft with names, as was the practice in Allied nations. On 14 December 1943 Ernst Heinkel Flugzeugwerke was asked if the proposed names of *Kormoran* or *Albatros* for the He 177 and *Marder* for the He 219 were acceptable.

Heinkel wanted *Herkules* and *Hermes*, respectively, however, the RLM could not agree, because it had been decided to use the names of predatory animals and birds. Little attention was paid to the fact that the crews of the He 219 had been calling it *Uhu* (Eagle Owl) for some time. Since no name subsequently appears in the files of the RLM or Ernst Heinkel Flugzeugwerke, it may be assumed that the effort to introduce an official name was a failure. At the start of the new year the status of the Ernst Heinkel Flugzeugwerke (EHF) was changed to Ernst Heinkel Aktiengesellschaft (EHAG), which was equivalent to an American stock corporation.

On the night of 21-22 January 1944 Bomber Command attacked Magdeburg.

One of the night fighter pilots who took off to intercept the raiders was Manfred Meurer, victor in 65 air combats with the British. Now *Kommandeur* of I./NJG 1, that night Hptm. Meurer was flying *Werknummer* 190070, a Heinkel He 219 A-0 coded G9 + BB. Twenty-two kilometers east of Magdeburg he collided with a Bomber Command Lancaster while attacking the enemy machine. The two aircraft went down locked together, and *Hauptmann* Meurer, *Ober-* *feldwebel* Gerhard Scheibe, and the seven crew members of the Lancaster lost their lives.

In January 1944 Meurer was succeeded by *Hauptmann* Förster. At forty-two years of age, Förster was certainly one of the "Grandpas" among German fighter pilots. He had joined the cadet corps at the age of fourteen, and after the First World War he fought as a member of a *Freikorps*. Förster then served in the *Reichswehr*, and

Summary of He 219 Prototypes as of 9 February 1944

V-No.	Werk.Nr.	Type of Testing	User	Remarks
V1	219 001	Stall tests, landing qualities	EHAG	Aircraft unsuitable for installation of other equipment due to different configuration
V2	219 002	Diving trials	EHAG	Destroyed
V3	219 003	Messerschmitt engine installation, undercarriage	EHAG	Aircraft sat about for a long time waiting for vibration-free MP8 propeller
V4	219 004	Engine test-bed	*E-Stelle Rechlin*	
V5	190 005	Weapons trials	*E-Stelle Rechlin*	Equipped with Rüstsatz II (MK 108)
V6	190 006	Catapult trials	*E-Stelle Rechlin*	
V7	190 007	Service trials	I. Jagdkorps	
V8	190 008	Undercarriage trials	*E-Stelle Rechlin*	
V9	190 009	Service trials	I. Jagdkorps	Destroyed
V10	190 010	Service trials	I. Jagdkorps	Destroyed
V11	190 011	Diving tests	EHAG	Replacement for V2, still under conversion to automatic catapult system and electrical horizontal stab. Adjustment. Because of special installations aircraft not used for testing other equipment.
V12	190 012	Service trials	I. Jagdkorps	
V13	190 052	Performance measurements	*E-Stelle Rechlin*	
V14	190 058	Increased range brake fittings tests with Dornier tail	EHAG	Currently under undercarriage emergency conversion to catapult
V15	190 064	GM-1, FuG 16 ZY, Uhu	*E-Stelle Rechlin*	Already with prototype installations. Further installations not possible for test reasons.
V16	still open	Jumo 222 and larger wing	EHAG	handling qualities and performance
V17	190 060	Service trials with DB 603 A and G supercharger	EHAG	Conversion to new intake scoop. After completion and calibration fllights to front-line unit for special missions.
V18	190 071	6 MK 103	*E-Stelle Tarnewitz*	Currently engaged in firing trials. To be delivered to E-Stelle when results known.
V19	still open	pressurized cockpit		

V20	still open	pressurized cockpit		
V21	still open	standard DB engine with internal flame damper		
V22	still open	standard DB engine with external flame damper		
V23	still open	Jumo 222, 2nd prototype		
V24	still open	2 x MK 103		
V25	190 122	single-core		currently being converted to single-core cabling destined for *E-Stelle Rechlin*
V26	190 120	oblique-mounted MK 108	I. Jagdkorps	To Tarnewitz after brief trials
V27	still open	B-1 series prototype with Jumo 222 E/F, larger wing, flattened canopy	*E-Stelle Rechlin*	
V28	190 068	Engine test-bed Period, speed brake Endurance trials	EHAG	No lengthy idle period allowable because of endurance trials
V29	190 069	De-icing	*E-Stelle*	de-icing section at Munich-Riem *Rechlin*
V30	190 101	Turbojet installation	*E-Stelle Rechlin*	Endurance trials
V31	190 106	Service trials	I. Jagdkorps	Numerous strength tests and measurements were canceled. A short 5-day program was carried out and then the aircraft was sent to the front.
V32	190 121	2nd GM-1 prototype	I. Jagdkorps	According to E-Stelle relocation of GM-1 in progress
V33	190 063	Antenna test-bed	Telefunken	Presently at Tarnewitz for air-firing. EHAG to install parabolic reflector, delivery to Telefunken.

The following aircraft, which went from the front to Lärz for conversion, were stationed at Rechlin and Lärz by the E-Stelle and the Rechlin/ Lärz test Detachment.

190 051	Endurance trials E.Kdo. Lärz	
190 055	Endurance trials E.Kdo. Lärz	
190 057	Endurance trials E.Kdo. Lärz	
190 059	Generator tests	*E-Stelle Rechlin* Abt. E-5
190 061	*E-Stelle Rechlin* Abt. E-2 testing of hot-water heating windscreen fitting tests	
190 062	Engine test-bed with testing of hot-water heating, Karcher heaters, oil coolers and radiators, automatic propeller pitch control	*E-Stelle Rechlin* Abt. E-3

The de Havilland Mosquito high-speed bomber. The "wooden wonder" operated at altitudes which German night fighters could not reach and remained invulnerable for a long time.

An He 219 A-0 on the ramp of the air base at Karup Grove, Denmark, in the summer of 1944. Their radar equipment removed, production aircraft were used here to train prospective Uhu pilots. No special training version with dual controls was ever built.

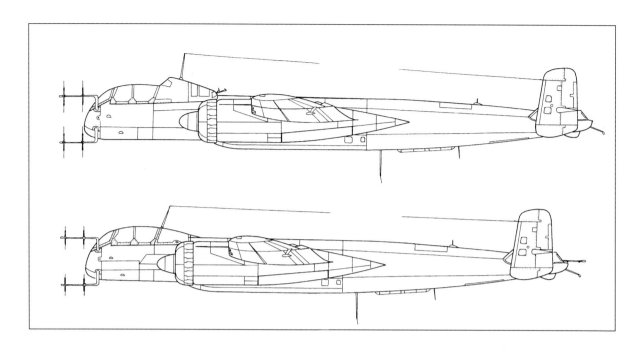

after his tour became a member of the reserve. Förster had himself reactivated in 1936 and was trained as a pilot. He received his baptism of fire while serving with ZG 1 and scored two victories during the campaign in western Europe in 1940. After he was shot down and wounded, Förster was assigned to the role of flying instructor and later served as a staff officer. In 1943 he retrained as a night fighter pilot, and on 1 June 1943 joined I./NJG 1. Förster's experience must have been rated highly, for he was chosen to test the He 219 in combat together with such notable night fighter pilots as Streib, Schoenert, Meurer, and Modrow. He scored six night victories prior to being named *Kommandeur* of I./ NJG 1, with which he added two more while flying the He 219.

A new night fighter unit, *Nachtjagdgruppe 10*, was formed in February 1944. Like *Jagdgruppe 10*, its sister day fighter unit, it was given the task of testing new tactics, armament, and equipment under combat conditions. The unit was equipped with the He 219 and the few Ta 154 *Moskito* night fighters to be built.

From April 1944 the RAF's mass attacks against the Reich slackened. This was partly due to the shockingly high losses it had sustained, but also to a shift in emphasis to raids on targets in northern France in preparation for the coming invasion. To compensate for this reduction in pressure on the German night fighter arm, the Royal Air Force increased the number of raids by De Havilland Mosquito high-speed bombers. That same month I./NJG 1 converted almost completely to the He 219. Based at Grove, Denmark, the *Nachtjagd-Ergänzungsgruppe* (Night-Fighter Replacement Training Group) received its first *Uhus* for the training of new crews. These machines were not equipped with airborne radar.

On 11 April 1944 the crew of an He 219 made the world's first successful ejection during an operational sortie. In recognition of this feat, Uffz. Herter and *Gefreiter* Perbix of 2./NJG 1 each received 1000 *Reichsmark* from Prof. Dr. Heinkel.

As a result of findings by the *Erprobungsstelle* (Experimental Station) and the RLM, the Heinkel Company was instructed to design a three-man cockpit for the He 219 and have it ready to fly by 31 May. He 219 A-0 *Werknummer* 190073 would be modified to test the new cockpit as the V19. To make room for the three-man cockpit, a 75-cm fuselage section would be inserted at Frame 8. Access to the cockpit would be aft of the nosewheel bay in front of the ventral weapons tray. Additional fuel tanks in the nacelle cones were planned for increased range. The pilot's ejection seat was retained, however, the radio/radar operator was provided with a swiveling seat and the gunner/observer with an adjustable folding seat for operating a rearward-firing MG 131. The series was to be built within He 219 A-5 production.

The RLM once again made plans to cease production of the He 219, this time because the front-line units did not want the aircraft! Heinkel went to the counterattack and interviewed operational He 219 crews. The following telex was sent to *Generaldirektor* Freydag on 14 April 1944:

Voices from the Front Speak about the He 219!

1. *Hauptmann* Förster, *Kommandeur* of the Nachtjagdgeschwader at Venlo.
Hauptmann Modrow, 2 victories.
Staffelkapitän Oberleutnant Baake, 3 victories.
Feldwebel Rauer, 1 victory on his first combat mission with the He 219.
2. Among the comments made by officers: Technical Officer *Oberleutnant* Hausdorf:

"Those crews who have flown the He 219 no longer want to fly any other night fighter."
Hauptmann Modrow:

"With any machine other than the He 219 I would not have been able to shoot down two fast four-engined bombers, the Me 110 would have been too slow. In six months the Me 110 won't be able to shoot down anything.

One can characterize flying the He 219 as old man's flying. The machine is very stable in the air, whereas the Me 110 must be trimmed constantly.

Landing the He 219 is childishly simple, even in bad weather. The entire *Staffel* is looking forward to converting to the He 219 in the near future."

Major Schönert, He 219 type specialist with the KDE, was equally fulsome in his praise of the He 219's speed and handling qualities.

General Schmidt and *Major* Müller-Trimmbusch, Galland's right-hand man, declared that the He 219 is the best machine extant and that it would be a mistake to kill the machine (remove it from the program). They do, however, want improvements, especially increased range and a three-man cockpit. A mockup of the latter has been inspected and approved.

3. Several days ago the ejection seats performed flawlessly in an emergency situation at Venlo. Equally successful were two landings with the nosewheel retracted, one at Venlo and one at Heidfeld. The aircraft did not flip over on landing, and damage to the machines was minimal.

4. *Major* Müller-Trimmbusch and other officers have made very unfavorable assessments of the Ta 154. With the present Jumo 211 it is too slow for the day fighter role, and its qualities make it poorly suited for the night fighter role. Single-engined performance is very poor, range is limited. No improvement in range or single-engined performance is expected from the Ta 154 with Jumo 213 engines.

The completed mockup of the future A-5 version was inspected and approved on 15 April. The aft-firing armament of one MG 131 Z*willing* (paired machine-gun) and the raised canopy for the navigator/radio/radar operator made necessary the addition of a strake for aerodynamic reasons. The designated prototype for this version was now to be the A-041 (190112), which would be designated the V34. The designation V19 for the A-073 was dropped, and

the planned installation of the three-man cockpit in the V19 was canceled.

On 21 April 1944 the British Air Ministry decided to commit long-range Mosquito night fighters over Germany in an effort to reduce bomber losses. If the Mosquito bombers had been a thorn in the side of the German night fighter arm, the Mosquito Mk. XIX night fighter armed with four 20-mm cannon and carrying A.I. Mk. VIII radar was to become a plague. From now on the Mosquito night fighters flew with the bomber stream with orders to locate and attack enemy night fighters, or carry out raids on their bases.

It was typical of the German situation that the He 219 was the only German night fighter capable of intercepting the marauding Mosquitoes. The type's performance was barely adequate, however. The DB 603 G engine was still not available. Some crews were growing tired of the long pursuits after the Mosquitoes and sought remedies. In some cases they had all camouflage paint removed from their *Uhus* and eliminated all excess weight in an effort to gain a few extra kilometers per hour. The mechanics did their part, keeping the engines finely tuned for peak performance. On 23 April 1944 a daylight raid by the American 15th Air Force halted production of the He 219 A-0 at Vienna-Schwechat. Part of a series of coordinated raids against aircraft industry targets in the south of Germany and Austria, it resulted in serious damage to the Heinkel production facilities.

In April-May 1./NJG 1 flew four ineffective missions against incursions by enemy bombers. After much urging, the crew of Nabrich and Habicht were issued an He 219 which had been cleaned up and lightened through the removal of four cannon and all armor. But apart from altitude sickness and fleeting encounters with their Mosquito quarry, the crew achieved noth-

ing apart from being fired at by German anti-aircraft guns. Then, on 6 May 1944, another crew was successful.

At five minutes before midnight, *Oberleutnant* Werner Baake, assisted by his radio/radar operator Uffz. Rolf Bettaque, shot down a Mosquito over Holland at an altitude of 8,000 meters. Baake was lucky, for the Mosquito was flying below its usual operating altitude and the Heinkel was able to dive on it from above and gain the airspeed necessary to overtake the "Mossie." But most of the Mosquito bombers continued to fly at altitudes of 9,000 to 10,000 meters where they could not be caught. The He 219's initial success could not hide the fact that production aircraft had proved incapable of matching the performance achieved by the V1. Nevertheless, in mid-1944 it was unquestionably the fastest and most heavily armed German night fighter, with altitude performance and takeoff and landing characteristics far superior to those of other types. Nevertheless, with full armament and almost full tanks, it could rarely reach altitudes in excess of 8,500 meters. It could only reach 10,000 meters if its fuel and ammunition were almost gone. In a normal instrument turn it lost 5 meters of altitude per second. Its previously calculated maximum speed of 605 km/h could only be reached without radar equipment. The maximum attainable speed with the FuG 220 and exhaust flame dampers installed was 560 km/h at an altitude of 6,200 meters.

The crew of *Feldwebel* Emil Heinzelmann and Uffz. Wilhelm Herling ran into trouble during a practice flight in G9 + FH, an He 219 A-0, *Werknummer* 190115 on 7 May 1944. The aircraft, which belonged to I./NJG 1, crashed near Sücheln at 0705 hours for no apparent reason. The recovery team, which reached the crash site the same day, found only wreckage. This was the beginning of a series of losses which was to continue until the end of the war. Inexperienced night fighter crews had to conduct familiarization flights in the morning or evening twilight, and were easy prey for the Allied fighters which constantly lurked about German air bases. Three more He 219 A-0s were lost this way on 21 May, 1 June, and 3 June.

On 13 May 1944 *General* Pelz testflew a new production aircraft at Vienna in order to form his own opinion of the He 219's performance. *General* Pelz was a first-rate pilot. He shut down one engine and then climbed to altitude. He then restarted the second engine and made a low pass over the field, followed by a rapid climb. Pelz made two takeoffs. After landing he said:

"The flight made a great impression on me. I had not thought that the machine was so good and that machines of this size class could be so different. I recently flew the Me 410 and there is a considerable difference. The He 219 is much easier to fly and has better handling characteristics. The He 219 flies almost as well on one engine

In-flight photo of A-2 *Werknummer* 290068. The camouflage scheme on the upper surfaces of the aircraft is interesting; instead of the usual irregular pattern of dark spots over a light base, it consists of a squiggle pattern of a light color over a dark base.

The same aircraft in night close-support finish on a transfer or training flight.

as most other types do on two. The He 219 provides adequate warning of an impending stall and dangerous flight attitudes, which has to be a significant advantage, especially at night. Takeoff and landing are extraordinarily simple. It is easier to land than the Me 410 and has the smoothness of the Ju 188. Since the quality of the new generation of pilots is constantly decreasing, good flying, takeoff, and landing characteristics are decisive in choosing an aircraft type."

The possibility of using the He 219 as a makeshift bomber interested Pelz very much, however, he was unable to make any sort of decision on this without the head of development, *Oberstleutnant* Knemeyer, who failed to show up. Pelz wanted the aircraft to be able to carry 1,000 kg, 500 kg, and 50 kg (incendiary) bombs. The Heinkel company requested that Pelz inform those

at the highest level of his good opinion of the aircraft when the opportunity arose.

II./NJG 1, which had been based at Arnhem/Deelen since the end of April, was now supposed to reequip on the He 219. First *I. Gruppe* at Venlo provided three or four aircraft for familiarization and study. Engineers from Heinkel and the *E-Stelle Rechlin* trained the aircraft mechanics. First to fly the machine was *II. Gruppe*'s technical officer, Lt. Fries. He carried out a flight to Vienna-Schwechat to clear up the question of spare parts with the Heinkel Company.

Lt. Fries took off on his first mission in the He 219 on 19 May 1944. The fighters were recalled when it was realized that it was only several Mosquitoes dropping Window to simulate a large bomber stream. While heading back to Deelen at 7,000 meters Lt. Fries' He 219 was detect-

ed by an enemy night fighter, which attacked near s'Hertogenbosch. The starboard engine caught fire. Fries shut it down and jettisoned the canopy as a precaution. As he pulled the lever he felt something strike his body and lost consciousness. When he came to the machine had fallen to 2,500 meters. Fw. Staffer, Fries' radio/radar operator, had already ejected. Blinded by blood from a gash on his forehead, Fries decided to abandon the aircraft. A new technical officer was appointed for *II. Gruppe* while Fries and his radio/radar operator lay in hospital. Both men recovered from their injuries, and when they returned to Deelen they found that they had been transferred to I./NJG 1 on account of their experience with the He 219.

The Allies had already raided the southern Heinkel factories several times. An attack by 194 B-24s of the 15th Air Force missed the factories and hit residential areas in the districts of Zwölfaxing and Schwechat. The next raid by the Americans on 8 April 1944 also produced unsatisfactory results. It was a foggy, day and the navigators were unable to locate the target.

On Sunday, 23 April, the USAAF struck a major blow at the aircraft industry in the area around Vienna. A force of 956 bombers from Italy crossed the Alps, and in the space of fifteen minutes dropped 292 tons of bombs of various calibers on the Heidfeld factory. Ninety-four workers died

in the rubble. Bombs fell again one day later, this time on Zwölfaxing.

During a conference at Berchtesgaden on 25 May 1944 *Reichsmarschall* Göring ordered further development and construction of the He 219 stopped. It was to be replaced as a night fighter by the Ju 388. The raw materials situation was given as the reason for canceling the He 219. Heinkel immediately retook the offensive and prepared a several-page memorandum, which it was hoped would change the minds of the RLM and the *Luftwaffe* command. At another conference at Rechlin on 13 June 1944 the decision of 24 May was overturned. This step was not taken without lengthy debate, however. *Herr* Sauer, leader of the *Jägerstab* (Fighter Staff), asked why the He 219 should be canceled given its success. Stabs.Ing. Baist of the *E-Stelle Rechlin* thereupon declared that the frontline units preferred the Ju 388. Heinkel's Dr. Freydag responded by stating that the Ju 388 had not yet reached the front-line units!

Why a staff engineer of the *E-Stelle Rechlin* should make a totally false statement to the leader of the Fighter Staff, even though his position meant that he must know the true facts, leads to some unusual conclusions. In this context it should also be mentioned that test pilots conducting comparison flights for the RLM were instructed, probably by senior officials, to fly

The Ju 388 V2, prototype of the Ju 388 J night fighter version, did not make its first flight until the end of January 1944. According to the RLM it was supposed to replace the He 219 on the production lines beginning in May.

93

the Heinkel competitor significantly more aggressively. The resulting impression had to lead to negative decisions. The desired follow-up contract and work on further development hung in the balance for a long time. Then the order was given to immediately reequip the He 219 with the DB 603 AS engine with improved superchargers for use in the anti-Mosquito role. The purpose of the memo was immediately overtaken, and it was not acted upon.

In the preceding months the Heinkel company, at the urging of Prof. Dr. Heinkel, had undertaken a careful check of the He 219's combat readiness at the front. The causes of unserviceabilities were to be investigated at once, and the potential for quick remedies examined. As a result of extensive cooperation between the Heinkel engineers and workers and the front-line maintenance facilities, the initial operational readiness of 20 to 25 percent was increased to 70 to 80 percent with extraordinary speed. Nevertheless, the survey revealed that only about half of the serviceable aircraft saw action due to a shortage of crews. Following on the heels of a meeting with personnel of I./NJG 1 for the purpose of hearing their complaints on 1 June 1944, Weber, head of the Heinkel check team, submitted the following report:

Shortcomings Reported by the Front-Line Units

Unintentional retraction of the undercarriage.

Determination of the cause was still not possible. Herr Schneider is remaining in Venlo and will try to determine the cause by examining the aircraft involved in the accidents.

1. Increased tire usage.

400 tubes and tires were changed in the course of 12,000 takeoffs. That means that a complete tire change must be made after every 12th takeoff. Nosewheel tire attrition is insignificant.

2. Burning out of brake linings and brake backing plates.

Brake wheel attrition in 12,000 takeoffs was 146 and 150 oil hydraulic brakes. The rapid wearing out of brake linings came in to particular criticism from the crews, as did the fracturing of brake shoes and the burning brakes. In several cases fire extinguishers had to be employed during night operations. Recently the unit has heard from Herr Viet of the *E-Stelle Rechlin*, who proposed the use of Me 262 brake wheels with double the heat absorption. Herr Schneider is familiar with these brakes, and he advised that several test brakes were already on hand in Vienna, but that they had not yet been installed and therefore no test results were available. We took the opportunity to suggest the following:

a) Apply brakes only after the aircraft has rolled a certain distance, by which time the machine's resistance had lowered its airspeed.

b) Do not apply brakes continuously, instead interrupt braking now and then.

c) Do not brake unnecessarily during training and use the entire airfield.

d) Watch to ensure that brakes do not become contaminated by oil, placing the entire braking load on the good brake.

3. Excessively high oil temperatures.

Oil coolers have proved inadequate during a continuous climb to 6,000 meters in pursuit of Mosquitoes. The cooling gills are in the full open position by 3,000 meters. Oil and water temperatures climb to more than 100° and the machine's climbing ability decreases. Missions often have to be aborted because of excessively high temperatures. The unit admits to having reset the thermostats to increase sensitivity and cause the gills to open sooner, however, no improvement was achieved. Attempts were also made to improve the thermostats by shortening the gills.

4. Oil Cooler and Radiator Failures

The unit characterized the failure of oil coolers and radiators as unbearable. In the past three months 102 radiators and 92 oil coolers had to be replaced on account of leaks.

5. Obliquely-Mounted MG 151 Cannon

The unit has now completed an oblique weapons installation of its own design and is flying with it. Installation of the first prototype took 14 days. One specialist and two assistants. The weapons were shortened by 24 cm, as a result of which performance was supposedly improved.

Reduced muzzle velocity. In this context, complaints were registered about the lengthy conversion time of the machines at Ludwigslust to the MK 108, which has been going on for five weeks now. The removal of machines from operations for such a period is considered unacceptable. The advantage of the Venlo installation is that there is no need to rearrange fuselage components. Note: In Rostock it was stated that the first machine completed at Ludwigslust is being picked up by Venlo.

6. Restraining Cable for Folding Hood

In 15 cases the folding hood restraining cable has failed. The unit desires stronger cables and a shortening of the cable guide tube so that the cable can be inserted and taken out without having to remove the rear fixed portion of the canopy.

7. Hydraulic Lines in the Nosewheel Bay.

On several aircraft the hydraulic lines on the left side of the fuselage in the nosewheel bay had to be replaced. The nosewheel had chafed against the line while in the retracted position. The lines have been rerouted in aircraft on the production line.

8. Fuel Return Line.

In two cases the fuel return line has broken off where it attaches to the breather, as a result of which all of the excess gaso-

line has run into the engine compartment. The pipe has already been replaced with a flexible pipe at the factory.

9. Rubber Seals.

The rubber seals used where lines join are unreliable and must constantly be replaced on account of leaks in the system. In particular, connections at the fuel pumps and tubing associated with the ejection seat system. During replacement it was repeatedly found that the rubber seals were completely destroyed. The unit desires metallic or fiber seals.

10. Flame Dampers with Attachment of the Collector Pipes to the Exhaust Stacks.

The unit advised that so far no problems have been encountered in flight with the new Klatte system, apart from poor mounting of the apertured partition.

11. Unsecured Couplings.

Technical officer *Oberleutnant* Hausdorf has the same concerns that problems might be encountered on operations, and therefore has the couplings checked before missions. Paragraph 4, Item d. of Circular Memo FB No. 68/44 VIII again calls for threaded pipe fittings to be secured with locking wire.

12. Hot Water Heating System.

The hot water heating system is unreliable due to sticking of the pilot valve on the firewall. The Vienna test unit also makes frequent complaints about the action of the pilot valve. The feed travel is too great, so that when the lever is at full deflection the force is not great enough to hold the valve. The valve also jams because of the influence of temperature. It is suggested that the spring's tension moment be reset or that two cables be used to activate the lever in each direction.

13. Elimination of the Armored Shield.

Oblt. Hausdorf advises that, in the opinion of all operational crews, the fold-

ing armored shield in front of the armorglass can be dispensed with.

Visit to Dehel:

The technical officer showed us the installation of a ventral gun position in the area of the aft entry hatch. Dehel used the existing MG 81 Z system (as used in the Ju 88). So far they have only completed the tubular framework with the weapon, and static attachment to the aft fuselage has been put off until the arrival of the statistician. The gentlemen are no longer convinced of the position's effectiveness, on the one hand because of the relatively shallow angle of fire above (threat of being shot down by a long-range fighter while landing), and because of the relative ineffectiveness of the light weapon. Furthermore, doubts have been raised about the installation's effects on the center of gravity and the compass system.

At the beginning of June the RAF formed two squadrons on the Mosquito N.F. Mk. XIX for use over the Reich. The British night fighters were equipped with improved A.I. Mark X airborne radar. With a range of 11 km, it was capable of scanning a much larger area than previous systems. The Mosquitoes were also equipped with "Monica" tail warning radar.

In a numerical coincidence, No. 219 Squadron opened its score against the *Luftwaffe* by shooting down an He 219. Details are sparse, but during their mission on 3 June 1944 Pilot Officer D.T. Hull, and his radar operator, Pilot Officer P.J. Cowgill of B Flight, shot down an unidentified German aircraft. Their Mosquito was an N.F. Mk. XVII, serial HK 248. After the intelligence officer had pored over the latest identification manuals for hours, he came to the conclusion that the twin-engined machine must have been an *Uhu*. The victory was not confirmed until after the war. The He 219 was certainly aircraft G9 + BL of 3./NJG 1, which crashed near Wilhelminsdorp with Obfw. Heinz Gall still on board. The aircraft's pilot, Hptm. Heinz Eicke, was able to parachute to safety.

On 7 June 1944 Hptm. Lüdtke and Obfw. Breer of *Nachtjagdgruppe 10* based at Werneuchen carried out the first local flight by the He 219 V34 (three-man cockpit). The pilot subsequently remarked:

"No noticeable differences were noted in takeoff, flight and landing characteris-

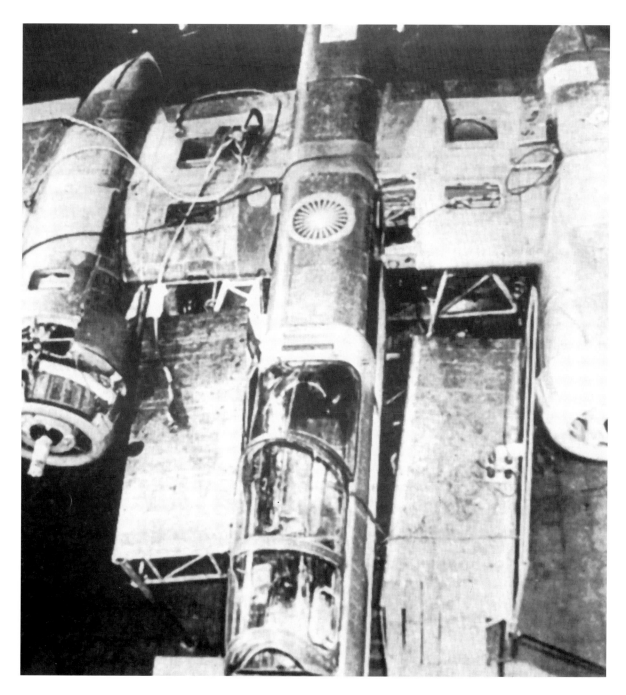

tics compared to the old model (two-man cockpit). On the contrary, I gained the impression that the machine was more stable in steep turns. During landing I had the feeling that the machine was softer in the landing process. I noted no changes whatsoever in longitudinal stability when the flaps were lowered or during the flare. Summing up, I believe that the three-man cockpit is a worthwhile development. My

radio/radar operator said, '*Herr Hauptmann*, sitting in this aircraft is like being in an automobile.'..."

It is 10 June 1944. High above the Zuider Zee at 9,700 meters an He 219 is on patrol. At the controls is O*berleutnant* Josef Nabrich, Uffz. Fritz Habicht is in the rear seat. Ground controllers report the approach of a formation of Mosquitoes, slightly lower than the *Uhu*. So far during

View from above of an He 219 A on the assembly line. The circular shape in the middle of the fuselage is the suppressed antenna of the Peilgerät 6 (D/F). It was incorporated into the cover panel for the forward fuselage fuel tank and was covered by a plexiglas panel.

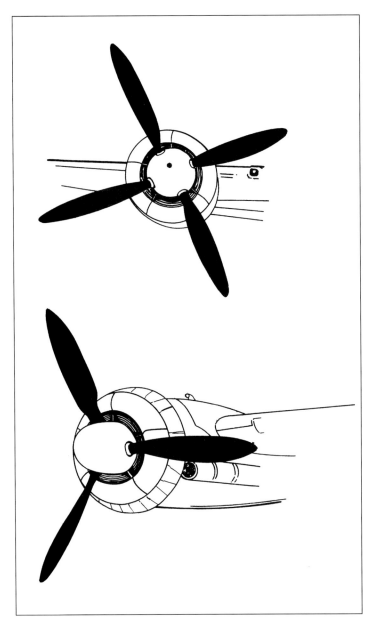

two wing cannon. The enemy machine's starboard engine immediately bursts into flames, and the Mosquito goes down in a spiral, out of control. After a few minutes Nabrich prepares to attack again, but then the Mosquito's bomb load explodes. The force of the blast causes the Heinkel to stall, and the pilot only recovers just above the cloud layer. Only a few scattered pieces of the Mosquito are found. The English crew later states that they both bailed out immediately after the first attack. They believed that they had been shot down by some new weapon, but not by a night fighter.

The *Uhu* and its crew were successful again the following night. The Mosquitoes had a long, difficult flight to Berlin. The slight height advantage that had helped them in earlier battles was now almost gone. West of Salzwedel Habicht spotted a Mosquito, so near that he was able to pick it up visually. Two bursts from the 20-mm cannon and the RAF bomber went down vertically. Seconds later a flash lit up the night sky—confirmation for the German crew. The lengthy pursuit at full power was too much for the port engine. Seconds later the DB 603 seized. With one propeller feathered, Nabrich brought the *Uhu* in for a single-engine landing at Perleburg. The Commander-in-Chief of the *Luftwaffe* sent the crew a congratulatory telegram and a gift of several bottles.

The Heinkel factories were kept busy trying to respond to requests and complaints from the units, producing equipment sets, and even special modifications to meet operational requirements. First the units wanted heavy caliber 30-mm guns, then they clamored for lighter 20-mm weapons.

II./NJG 1 at Dehel had been flying several He 219s since June 1944. That unit's pilots were used to the Bf 110 and had a hard time getting used to the *Uhu*. They did not feel that the He 219's perfor-

the month of May Nabrich and Habicht have been successful four times while patrolling the bomber route to Berlin in their lightened machine. On this night the sky is once again a dangerous place. The possibility of being shot down by friendly flak or an enemy Mosquito is ever present. Habicht, his face pressed to the visor of his radar display, directs his pilot to reverse course for about ten and a half kilometers. A Mosquito is flying east at high speed. Over Osnabrück they close to within firing range. Nabrich fires a short burst from his

mance advantage over the Messerschmitt was convincing enough, were extremely uncomfortable with the pilot's position ahead of the propellers, and were particularly critical of the *Uhu*'s two-man crew. Unaware of Heinkel's efforts to develop a three-man cockpit, on several machines the aft entry hatch, which was intended to provide access to obliquely-mounted guns, was replaced with a plexiglass panel, through which a third crew member could visually scan the area beneath the fighter. They even installed an MG 81 Z in this makeshift ventral position for defense against the Mosquitoes.

Since the *Uhu*'s performance advantage over the Mosquito was slight, a special Mosquito hunter version was planned. Designated the He 219 A-6, it was to have all engine and ammunition tank armor removed, along with the oblique-armament. By the time the first of these Mosquito hunters was delivered, however, use of the He 219 against the Mosquito had officially been banned. The following letter was written by EHAF on this topic:

"Subject Night Interception of Mosquitoes.

*After initial success with the He 219 against Mosquito night incursions, in recent weeks night missions against Mosqui-*toes *have been fruitless. In many cases contact with the enemy was made or targets were acquired on the Lichtenstein, however the targets suddenly gained speed when the night fighter had worked its way to within 1 to 2 kilometers. The speed of the enemy aircraft then became so great that they could not be caught at maximum power. This means that the enemy now has a tail warning radar or, if he already had the tail warning radar, that the Mosquito has received more powerful engines (Griffon).*

According to our information, the Griffon-powered Mosquito has a maximum speed of about 700 km/h at an altitude of approximately 8,000 m. Our night fighters, therefore, require a significantly better performance than before. The units are demanding a speed increase of 60 to 100 km/h for the He 219. Since, according to the Dornier Company and our own opinion, the first prototype of the night fighter Do 335 will not fly before spring 1945 and initial Do 335 production will be dedicated to the needs of the day fighters, the Do 335 night fighter cannot be expected to enter service before the fall of the year. As the fastest aircraft available, the He 219 is the only suitable aircraft for intercepting the Mosquito. The aircraft's speed absolutely must be increased to cover the necessary performance range at least until next autumn.

For this reason, we have for months been suggesting the installation of the Jumo 222 in the He 219. The entire Luftwaffe has no interest in this engine. It was not until our proposed combination of this engine and the He 219 resulted in extraordinary performance estimates (700 km/h) that the Ju 388 was switched to the Jumo 222. Simultaneously, all production of the Jumo 222 (beginning in spring 1945 at 250 examples per month) was allocated to the Ju 388 (see delivery program...).

Based on the new tactical situation resulting from the increase in the Mosquito's speed, we consider this decision wrong, since it is acknowledged that the Ju 388's performance is inferior to that of the He 219, and to date it has not completed manufacturer or service trials.

The He 219 night fighter equipped with the Jumo 222 flew several days ago in Vienna. Handling characteristics are essentially good. Performance has not yet been measured, however, airspeed indications in level flight indicate that the performance figures predicted by us will be achieved. The He 219 will thus reach 666 km/h at an altitude of 10 kilometers while equipped with flame dampers and all antennas, including the Lichtenstein antenna. Use of methanol-water will allow 700 km/h to be reached at an altitude of 8 kilometers. We intend to complete five Jumo 222-powered He 219s in the Vienna development works by the end of this year. The contract for this has been issued by C E2. No Jumo engines have been allocated for September, however. Furthermore, there is no allocation of Jumo engines from next year's production for converting some of the He 219 A series to Jumos.

In an effort to produce faster night fighters by the end of the year, we have run tests with the installation of the Jumo 213 E/F in the He 219. This would result in a maximum speed of 606 km/h at an altitude

of 10 kilometers. A speed of 640 km/h at an altitude of 8 km will be possible with the use of methanol-water. Compared to the He 219s now at the front powered by the DB 605 AA, this represents a speed increase of 45 km/h. If we can be allocated Jumo 213 E/F engines in September-October, from November we will be in a position to convert three He 219s to the Jumo 213 E/F on the production line per month. Our efforts to produce a significantly faster night fighter this year depend upon the capacity of our prototype construction program. Slightly faster night fighters are always better than nothing. But in our opinion something must also be done to exert extraordinary pressure for an immediate production program for the He 219 in the variants described above.

The company began designing and building the He 219 B, a heavy high-altitude night fighter based on the He 219 intended to meet the future threat of high-altitude heavy bombers. In all the company's efforts, it was apparently overlooked that, because of the general raw materials shortage, the interval between a request from the front and its realization by the factory was growing ever wider. Extraordinarily confusing, not just for the present-day historian, but also for the technicians and supply personnel in the units and headquarters at the time, was the multitude of *Rüstsatz* (equipment set) designations which inexplicably changed with each variant of the He 219. Instead of concentrating on increased production of the proven night fighter, Heinkel continued working on a variety of projects developed from the basic He 219.

In production in June were the A-2 and A-5 versions, which were essentially similar, apart from power plants. In preparation were:

• the A-6, a radically lightened Mosquito hunter

- the three-man B-1 night fighter powered by the Jumo 222
- the B-2 high-altitude night fighter powered by the Jumo 222 with exhaust-driven turbo supercharger
- the four-man C-1 night fighter with Jumo 222 engines and manned four-gun tail turret
- C-1 fighter-bomber with four-gun tail turret
- Hü 211 high-altitude reconnaissance aircraft.

Crews attacking with the obliquely-mounted MK 108 cannon faced the constant threat of flying wreckage damaging their own machines, and pilots often had the upwards-firing weapons removed or replaced with the proven MG FF/M. The units declared that if the MK 108s were to be retained, an entirely different firing angle would have to be found. Either the angle had to be smaller, so that the pilot could open fire from a greater range, or so great, almost 90°, that he could fire from a position almost directly beneath the bomber. In either case, the attacking fighter would not be struck by pieces flying from the target aircraft.

The weapons manufacturer Mauser subsequently developed a mount for the MK 108 designated the L 188, which would allow the pilot to set the weapons at any angle between 45 and 85 degrees. This weapons arrangement, which Ober*leutnant* Welter later demanded for the Me 262, never entered front-line service.

Although the MK 108 was in short supply and the night fighter pilots had no use for it, all He 219s continued to be delivered with these weapons installed as upwards-firing armament, which was immediately removed by front-line workshops.

During the night of 15-16 June 1944 2./NJG 1 lost He 219 A-0 G9 + RK (*Werknummer* 190180), which crashed one kilometer north of Leersum. The crew (Uffz. Willi Beyer and Obgefr. Horst Walter) was killed.

At about this time the *Kommandeur* of I./NJG 1, Förster, was promoted to the rank of *Major*. The He 219 that he flew from Venlo was listed in the operational readiness reports as an A-0, but it is a certainty that it was not a standard type. The air intakes on both sides of the engines were the "shark mouth" type used by the A-5. The air intakes for the wing heating system, located on the starboard wing right next to the landing light, were totally absent. Close examination of photographs of the *Kommandeurmaschine* reveal an overpainted area over the inlet for the cockpit heating. This was most likely the original site of the centrally-mounted FuG 212 antenna, which was deleted after the C-version of the SN-2 was installed. The painted guide line from the hand holds to the foot step on the port side of the forward fuselage was probably dark red. Several machines had rubber pads glued on which could be felt with the tip of the foot in the darkness. The exteriors of the engine nacelles were completely soiled in the area of the main undercarriage wells and not painted black. This phenomenon was seen on other aircraft powered by the DB 603, for example, the Do 217 M and N and the Me 410.

After a visit to I./NJG 1, Heinkel employee Hilber sent a list of successes by the He 219 to Prof. Dr. Heinkel as an attachment to his trip report (No. 779/44). It included all air victories scored with the He 219 until 16 June 1944. In order to avoid repetition, however, all known victories until 3 November 1944 are included in the following table:

Name	Date	Number of victories on day of mission	Total
Obstlt. Streib	12/06/44	5	5
Hptm. Frank	26/07/43	2	
	23/08/43	1	
	31/08/43	3	
	06/09/43	1	7
Hptm. Meurer	18/10/43	1	
	22/10/43	1	
	03/11/43	1	
	21/01/44	2	5
Hptm. Modrow	30/03/44	2	
	22/04/44	3	
	24/04/44	3	
	01/05/44	1	
	11/05/44	2	
	12/05/44	2	
	21/05/44	1	
	22/05/44	1	
	28/05/44	3	
	11/06/44	1 Mosquito	
	13/06/44	1	
	21/07/44	1	22
Hptm. Förster	21/05/44	1	
	16/06/44	1	2
Hptm. Strüning	10/05/44	1	
	12/05/44	1	
	21/05/44	1	
	22/05/44	1	
	24/05/44	2	
	03/06/44	1	
	06/06/44	1	
	16/06/44	2	
	18/06/44	1 Mosquito	
	21/06/44	4	
	18/07/44	1 Mosquito	
	19/07/44	1 Mosquito	17
Oblt. Baake	11/04/44	1	
	24/04/44	2	
	06/05/44	1 Mosquito	
	11/05/44	1	
	22/05/44	1	6
Maj. Karlewski	22/04/44	1	
	21/05/44	1	
	22/05/44	1	
	24/05/44	1	
	10/07/44	1 Mosquito	
	29/07/44	1	6
Oblt. Nabrich	24/03/44	1	
	10/05/44	1	
	12/05/44	1	
	24/05/44	2	
	11/06/44	1 Mosquito	
	12/06/44	1 Mosquito	
	16/06/44	2	
	13/08/44	1	10

Oblt. Henseler	27/04/44	2	
	21/05/44	1	
	22/05/44	1	
	24/05/44	1	5
Oblt. Schön	23/09/43	1	
	23/09/44	1	2
Lt. Hittler	27/04/44	1	
	03/06/44	1	
	13/06/44	1	3
Fw. Rauer	11/04/44	1	
	27/05/44	1 Mosquito	
	28/05/44	1	3
Fw. Morlock	27/05/44	1	
	28/05/44	1	
	31/05/44	1	
	16/06/44	2	
	24/07/44	1	
	09/10/44	1	
	03/11/44	6 (+ 1 unconfirmed)	14 (15)
Uffz. Wildhagen	24/04/44	1	1
Obfw. Stroehlein	26/04/44	1	
	06/07/44	1	2
Oppermann	16/06/44	1	1
Bane	06/06/44	1	1
Schäfer	16/06/44	1	
	29/07/44	1	2
Oblt. Finke	02/07/44	1 Mosquito	1
Lt. Fries	29/07/44	1	1
Uffz. Frankenhauser	13/08/44	1	1
Oblt. Thurner	02/11/44	1	1
		133	133

Following a successful attack on Zwölfaxing by the USAAF on 8 July 1944, which caused Aircraft Halls II and IV and the He 177 prototype shop to go up in flames and cratered the airfield, Heinkel abandoned the factory and assembled its remaining capacity at the Heidfeld night fighter factory, which had also been damaged. As a result, plans to construct the He 219 there had to be abandoned.

On 11 July 1944 the RLM sent a priority telegram to Schwechat to accelerate the replacement of the He 219 A-2 by the A-5. The first five A-2s, which were produced in Vienna, had been completed in September 1943. Since then there had been continuous problems with the Uhu's fuel delivery system. The A-5, which incorporated a reworked system, was supposed to eliminate these problems.

The *Luftwaffe* received its first He 219 A-5 from Schwechat in March 1944 and from Marienehe in the following month. Various *Rüstsätze* were available for use with the A-2:

R1 two 30-mm MK 108 in the fuselage tray
R2 two 20-mm MG 151/20
R3 two 30-mm MK 103 in the fuselage tray
R4 three-seat model (later designation A-5 with sets R1 – R3)

All versions of the A-2 were equipped with a standard armament of two cannon in the wing roots and *"Schräge Musik"* (obliquely-mounted upwards-firing guns).

On 30 July 1944 Gen.Dir. Frydag decided that the three-man cockpit (flat top, navigator/radio/radar operator in front, radar operator behind) was to be incorporated into the A-5 series. A revised arrangement was to be adopted at a later date, with

the navigator/radio/radar operator in the rear and the radar operator in front. For comparison purposes, the completed three-man cockpit was to be available by mid-September at the latest. Nevertheless, the He 219 A-2 remained in production until at the beginning of 1945.

The development and testing of the He 219's ejection seat system was essentially concluded in July 1944. For the crews, safe ejection was now assured at the following speeds and altitudes:

max. km/h 550 at altitude of 1,000 meters
max. km/h 620 at altitude of 4,000 meters
max. km/h 780 at altitude of 8,000 meters

On 25 July the RLM issued a contract for the construction of twenty prototypes of the He 219 powered by the Jumo 213 E/F. The power plants were obtained at the expense of the Ta 152 H, for which the day fighter arm was waiting at least as urgently as the night fighter arm was the He 219. The new engines were expected to boost maximum speed to 640 km/h at an altitude of 9,500 meters. The only possible Heinkel competition for the Dornier Do 335/435 was a projected version of the He 219 powered by the Jumo 222 E/F. Forecast maximum speed was 700 km/h at 10,000 meters. Five prototypes were already under construction and were supposed to fly by October 1944; however, even Heinkel did

not know where the power plants were supposed to come from. According to a message from the RLM GL/C/B2, the DB 603 AA power plant was no longer adequate for production of the He 219 A-2. Effective immediately, this version was to be equipped with the DB 603 E, resulting in the designation He 219 A-7.

A total of 105 He 219 A-0 to A-5 aircraft were built by the last day of July 1944. This total includes fifteen A-0s from the Rostock factory. Production of the He 219 A-2 at Vienna-Schwechat, five aircraft per month at first, finally got under way. I./ NJG 1, which had officially received 36 *Uhus*, had just a dozen of these aircraft on strength, of which seven were serviceable. II./NJG 1 was supposed to have nine A-0s, however, it was still awaiting delivery of these, and conversion to the type was yet to begin. *IV. Gruppe* was in a similar situation. NJGr. 10's official complement was 12 *Uhus*. In reality it had just fifty percent of this figure, and of these only three were serviceable. On 14 July an He 219 crashed for unexplained reasons while on approach to land at Venlo, and one week later another went down in Schleswig after developing engine trouble.

Prototypes and several A-0 machines continued to fly as test aircraft. The parent company was still using the V11, an "old

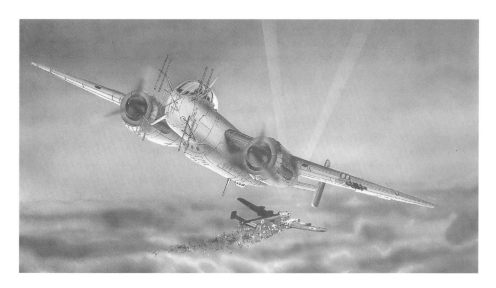

timer," for general handling tests. The V21 and V22 were used to test engine deicers and flame dampers. In July the V16 was flying with Jumo 222 B engines and also had a greater wingspan as planned for the B-1. The V23 was also converted to take the Jumo 222 B. The *E-Stelle Rechlin* began testing the deicing system using the V29. The twelfth pre-production machine, the A-012 (*Werknummer* 190062, RL + AB), flew with experimental flame dampers and a new Kärcher heating system. The V30 was equipped with a BMW 109-003 turbojet engine bolted on beneath the fuselage, a combination which was supposed to put an end to the Mosquito's speed advantage once and for all. Initial tests were unsatisfactory, however. The turbojet engine provided an enormous increase in acceleration, however, top speed was improved little. When the BMW 003 was shut down in flight there was a serious loss of performance because of increased drag. Further experiments were carried out, however, and the combination was probably tested in action. The V15 and V25 were fitted with various types of radio equipment. At that time the technicians at Telefunken were striving to install the FuG 16 ZY as a VHF voice radio and fighter control system. Also planned was the installation of the *Morgenstern* antenna for the SN-2 and the

FuG 218 and *Berlin* N 1 with parabolic reflector dish. The two test-beds were based at the Telefunken company airfield at Diepensee near Berlin. Already in October 1943 Ing. Kösel had installed and tested the FuG 350 *Naxos* (H2S homer) and FuG 120 *Bernhardine* (ground to air communications, information—bearing of ground station and commentary on air situation—transmitted to aircraft, received in code on strip of paper) in the He 219.

The veteran V5 was used up in oblique weapons tests with the MK 108 at Tarnewitz. Examinations revealed damage to the fuselage structure caused by the recoil of the 30-mm weapons.

The V33 flew for Telefunken and carried various radar systems.

The V34 was used as the test-bed for the three-man cockpit. Two other machines were converted for further tests with this cockpit, WNr. 190042 and the V23, which had been earmarked for the *Erprobungsgruppe Werneuchen*.

The V4 and V6 were used for the ejection seat trials described earlier. This was a field in which the Ernst Heinkel AG undoubtedly conducted groundbreaking work. The *Uhu* was the first production aircraft in the world to be equipped with ejection seats and apart from the He 162, which entered service later in the war, was the

only combat aircraft used by any of the participants to be equipped with this life-saving device. Not until six years later, during the Korean War, did other nations use comparable equipment.

Given the large numbers of foreign and forced workers employed in Heinkel's Rostock facility, it is not surprising that the Allies were constantly kept informed about He 219 production. Based on photographs taken by high-flying RAF Mosquitoes, another attack on Marienehe was planned.

On 4 August 1944 a force of 117 four-engined bombers of the American 8th Air Force took off from bases in England. The next attack took place on 25 August. Altogether 676 tons of bombs fell on the expansive factory grounds in the two raids. The resulting destruction certainly had an impact on the already halting *Uhu* production program. The frequent attacks and resulting materials shortages forced Heinkel to redesign major components for construction in wood. The Hütter firm, which had extensive experience with gliders, was

brought into the program. A high-altitude high-performance fighter based on the He 219, the Hü 211, was developed, and an initial batch of twenty was anticipated. These machines were supposed to be completed by spring 1945.

Venlo now faced an increased threat of Allied air attack, and on 5 September 1944 I./NJG 1 was transferred to Münster/Handorf. At 1750 hours, during a training flight to Rheine, two He 219s of 2. *Staffel* were attacked by American fighters. G9 + DK, an He 219 A-2, crashed with its crew near Hopsten at 1755. The second machine, an A-0, tried to escape under cover of the Hopsten air base's flak belt, however, it was hit while on short final and crashed onto the concrete runway at 1758.

The first loss at Münster/Handorf occurred on the night of 25-26 September. As a result of battle damage G9 + EK was unable to lower its landing flaps or undercarriage and came down too hard while attempting a belly landing. The crew were rescued with serious injuries.

1 October 1944 was a black day for I. *Gruppe. Major* Paul Förster, the unit's third *Kommandeur,* and his radio/radar operator Fritz Apel were killed when their aircraft crashed on short final while testing a new instrument landing system. Förster had twelve victories to his credit, ten of them scored at night.

At 0005 hours on 5 October a Mosquito N.F. Mk. XVII, serial number HK 245, piloted by Flying Officer G.S. Irving of No. 125 Squadron lifted off from the runway at Middle Wallop, Hampshire, and set course for the Channel Coast. After setting his pilot on course for radio beacon JQ, radar operator Flying Officer G.M. Millington switched on his Mark X radar.

Since the British had first used Window (strips of aluminum foil) to blind German radars in their raid on Hamburg eighteen months earlier, the *Luftwaffe* had tried with success to use its own version, called *Düppel.* If not impervious to *Düppel,* the latest A.I. radar carried by the Mosquito was at least less susceptible to its effects

than the earlier models. Suddenly, "Greengrocer," a fighter control station located near Brussels/Melsbroek, reported in. It instructed HK 245 to set course for radio beacon JP. "Greengrocer" was a secret night fighter command station 3,500 meters below the Mosquito in German-occupied territory. Served by trained British agents, the device was a large, portable trunk with foldable DF and transmitter antennas and a gasoline engine which drove a generator. It sent a constant stream of coded pulses into the air. High above in the night sky Millington checked his position—the Morse signal JP was coming in loud and clear. It was a dark night with 6/10 cloud, stratus at 2,100 meters, but clear enough at their operational altitude. "Greengrocer" passed them on to "Rejoice." This forward unit accompanied the Mosquito until it reached the reception area of "Milkway" station. For 45 minutes the British night fighter flew up and down its patrol route. Then, suddenly, "Milkway" reported that it had a second target on its screen, approximately three miles east of the Mosquito. Some miles away the sky was lit by fires from the burning city of Duisburg. Irving put the machine into a turn, made a 360 degree turn, and then returned to his course. Then Millington picked up a target on his scope. It was moving rapidly in the direction of Duisburg. After a few minutes it disappeared into a mass of targets—RAF bombers heading home after devastating Duisburg. At 0220 "Milkway" called out a new bogey at the Mosquito's eleven o'clock position. At a range of 3.5 miles a target appeared on Millington's radar. It was about 2,000 feet above them. The RAF night fighter made a 180 degree turn and gave chase at an indicated airspeed of 515 km/h. The enemy aircraft's pilot was obviously experienced, as he constantly changed course by almost 50 degrees and climbed and descended with no discernable pattern. Then the German turned back toward Duisburg. Fearing that they would lose this bogey, too, Irving advanced both throttle levers to the gate. The Mosquito's Merlins were now almost in the danger zone. The distance between the two aircraft was about 1,000 meters when suddenly two navigation lights came on, slightly to the east. Acting on this unexpected signal, Irving immediately corrected his course. At maximum speed in a gentle dive, he closed to within 400 meters. The Mosquito was now flying at 612 km/h and was several meters below the German. The target aircraft's white-green exhaust flames were now clearly visible. The crew of the Mosquito thought that it was a Do 217, however, the enemy aircraft's speed was much too great for that. Then the two large crosses on the undersides of the wings became visible. When the extended engine nacelles and twin tails became visible, Millington identified the machine as an He 219. Irving allowed the Mosquito to drop back 750 feet, raised the nose, and fired a two-second

burst from his four 20-mm cannon. The *Uhu*'s starboard DB 603 exploded. A second burst and the fuselage fuel tank caught fire. The Heinkel began to roll over slowly. Another burst exploded the starboard fuel tank. Circling over the stricken enemy machine, the Mosquito crew saw it hit the ground and explode.

During the night of 14-15 October an He 219 of 1./NJG 1 crashed near Münster after an engagement with several Lancasters. The crew of Uffz. Frankenhauser and Uffz. Biak were able to abandon the aircraft, however, the radio/radar operator sustained fatal injuries while bailing out.

One night later G9 + BH was lost eight kilometers northeast of Bremen following air combat. It was another machine of *1. Staffel*, this time *Werknummer* 290 002, an A-2. Pilot Oblt. Stieghorst and radio/radar operator Uffz. Frunske both went down with the aircraft.

A total of 187 He 219s had been completed by the last day of October 1944, 114 at Schwechat and 73 at Rostock. Many of the A-0 pre-production series *Uhus* also served as prototypes. The V41, at the time the latest prototype, carried out tests with the Jumo 213 E at Schwechat.

Several pilots flew outstandingly successful missions with their He 219s. One such was *Oberfeldwebel* Wilhelm Morlock. Assisted by his observer, Fw. Alfred Soika, on the night of 2-3 November he scored six certain and one probable victories in just 12 minutes. The next night, however, he fell prey to a Mosquito. His machine, an A-0 with the code G9 + HL, crashed onto a slope of the Teutoburg Forest, near Ibbenbüren. Morlock's radio/radar operator ejected from the machine and sustained only minor injuries.

On 27 November Oblt. Josef Nabrich, *Staffelkapitän* of 3./NJG 1, was killed in his vehicle during a strafing attack by British fighter-bombers on Reichsstrasse 54 from Handorf to Telgte.

The four-engined Lancaster bomber was Bomber Command's main weapon in its campaign against Germany.

Early on the morning of 28 November 1944, Lt. Kurt Heinz Fischer, a twenty-two-year-old pilot fresh from the night fighter school, took off on a training flight over Westmünsterland in an He 219 A-2 (G9 + MK). Behind him radio/radar operator Uffz. Hermann Bauer sat practicing with the radar set. Approximately thirty minutes earlier a formation of Tempest Vs had taken off from Volkel, Holland, on an armed reconnaissance mission. The formation leader, Flight Lieutenant A.F. Moore of No. 56 Squadron, led his aircraft in the direction of Münster. After strafing the barracks in the north of Münster, a worthwhile target appeared in the shape of a passenger train on the Münster-Rheine line. The fighters shot up the train near the town of Sprakel. While the attack was in process, the pilot of one of the British fighters sighted Fischer's Heinkel. The fast single-engine Tempests rapidly overhauled the *Uhu*. By the time Uffz. Bauer spotted the British fighters it was already too late. Moore fired burst after burst into the night fighter, which began to burn and lose altitude. Lt. Fischer radioed his nearby base that he was under attack and would attempt a forced

landing. He set course for a long field, however, while on approach the badly damaged machine suddenly rolled over and crashed into a wood just short of its intended landing place. Bauer ejected, but he died of injuries sustained when he struck the ground. Residents of the village of Reckenfeld who observed the accident attempted a rescue, but they were too late and Fischer died in the flames.

Flight Lieutenant Moore had little time to watch the crash, for at the same time, 0835 hours, he spotted another Heinkel. He immediately hauled his Tempest (US-B, serial EJ 536) into a turn, and with the others attacked the German aircraft. Pieces were seen to fly off the Uh*u*, but it refused to go down. Dwindling fuel forced the British fighters to break off their attack and head for home. Moore claimed the second machine damaged.

The night of 31 November-1 December was a fateful one for Uffz. Frankenhausen, who had been forced to bail out of his *Uhu* just six weeks earlier, and his radio/radar operator Uffz. Erwin Fabian. Their Heinkel (G9 + CH) suffered engine failure and crashed while on approach to land at Handorf. On 17 December No. 56 Squadron carried out another armed reconnaissance over Münster. Flight Lieutenants Ross and Shaw intercepted an He 219 which was on final approach to Handorf. The aircraft swung when it touched down on the runway and crashed into a building. During the night of 18 December a Mosquito N.F. MK.XIX of No. 157 Squadron RAF based at Swannington shot down an *Uhu* of II./NJG 1 near Osnabrück. On the next night *1. Staffel* again had contact with British long-range night fighters.

At about 2200 hours Uffz. Herbert Scheuerlein took off in G9 + GH, an He 219 A-0 *Werknummer* 190229, to intercept a reported incursion. After only fifteen minutes in the air, by which time the night fighter was in the Südlohn area, there were suddenly explosions in the wings and fuselage. With his radio/radar operator Uffz. Heinze showing no signs of life and the situation becoming more than dangerous, Scheuerlein jettisoned the canopy, pulled in his legs, and ejected from the aircraft. The Heinkel roared past beneath him and spun towards the ground. After quickly separating from his seat, he opened his parachute and made a relatively smooth landing. He felt a sharp pain in his neck—in his haste he had forgotten to unplug the headset wire, and the throat microphone had almost strangled him. This crash was the last of six *Uhus* lost by I./NJG 1 in December, and thus its last in 1944.

While all around the Reich the *Luftwaffe*'s ability to conduct operations was sinking steadily, the senior staffs and Heinkel continued to plan and design. That month the Hü 211 mockups built by Hütter in Ravensburg and Kirchheim/Teck were inspected by representatives of the RLM, Heinkel, and the *E-Stellen*. The first thirty machines were to have a fixed rearward-firing armament. Firing trials were carried out before the end of the year, and the design was finalized.

The files of the quartermaster of *6. Abteilung* reveal that I./NJG 1 was having major serviceability problems on 31 December 1944. All forty of its He 219s were in need of overhaul or maintenance.

NJGr. 10, by now reduced to the size of a *Staffel*, had just five He 219s and Ta 154s, of which four were serviceable. Ironically, there were now more He 219s available than trained crews to fly them. At Handorf on New Year's Day there were just eleven operational He 219 crews. The previous evening the *Staffelkapitän* of 3./NJG 1, *Oberleutnant* Oloff, and his radio/radar operator Fw. Fischer had been wounded and put out of action indefinitely when their He 219 A-2 came down near Schleiden after being hit by enemy fire.

Two other crews were only conditionally operational, and 21 others were waiting for conversion training on the *Uhu*.

As a result of delays with the Jumo 222, there was still no sign of the A-7 and the B-series high-altitude fighters, which had been promised for the summer of 1944. Now, in January 1945, production of the A-7 finally began, although with the DB 603 G. The A-7 prototypes, the V25-V27, had been tested in December. As usual, various armament combinations were offered, however, the leading crews now preferred *Rüstsatz 4*, which consisted of two MG 151/20s in the fuselage tray and two in the wing roots. Experience with the Bf 110 and Ju 88 had shown that this armament was sufficient for ace pilots to bring down bombers. In addition, the heavier armament proved a hindrance in the event of Mosquito attack.

On 1 January 1945 the *Luftwaffe*'s day fighter arm launched one of its most controversial operations—"Operation *Bodenplatte*." Its objective was to destroy the Allied air forces on the ground. The operation ended in disaster for the *Luftwaffe*, which lost 214 aircraft in four hours. Pilot casualties were 151 killed or missing. Sixty-three

pilots were captured, and 18 returned with wounds. Approximately forty percent of German losses were inflicted by friendly anti-aircraft fire. Machines of the night fighter arm participated as pilot aircraft. On 1 January 1945 II./NJG 1 lost three aircraft and their crews. During the night of 1-2 January 1945 an He 219 fell prey to a Mosquito Mk. XXX of No. 604 Squadron RAF. The aircraft was seen to go down near Mönchengladbach.

No. 157 Squadron RAF added to its victory list on the night of 5 January when one of its Mosquitoes located and shot down the He 219 flown by the crew of Ströhlein and Kenne. The aircraft crashed five kilometers south of Wesendorf. Radio/radar operator Kenne was able to eject safely. Another He 219 came down near Diepholz as a result of engine failure.

On 7 January 1./NJG 1 lost an He 219 near Bad Iburg. Then, one week later, *3. Staffel* lost an A-2 in air combat. The aircraft came down 9 kilometers north of Münster in the Sprakel/Gelmer area. Whether this was the same aircraft claimed by Flight Lieutenant Crafts of No. 274 Squadron RAF is uncertain. In any case, while Flying Tempest V EL 762 he attacked an He 219 near Hamm and saw it going down eight kilometers to the east of the town. If it was the same aircraft, then pilot Lt. Lehr must have regained control of his machine and flown 50 kilometers before crashing on approach to Handorf air base.

During the night of 16 January a *Uhu* was lost while returning from Cologne to Handorf. The machine crashed near Laggenbeck. The crew managed to escape.[*] I./NJG 1 lost one more He 219 by the end of January.

"Gemse" was the code-name of a German radio beacon located near the cities of Krefeld and Mönchengladbach. On the evening of 4 February about a half dozen *Uhus* of I./NJG 1 were circling over "Gemse" waiting for bombers reported heading for the Ruhr. At the controls of aircraft G9 + WH of *1. Staffel* was twenty-year-old Uffz. Karl Heinrich Thurow. Sitting behind him in his ejection seat was radio/radar operator Gefr. Neff, twenty-two years old. Thurow had been maintaining an altitude of about 8,500 meters for about 15 minutes, constantly changing course,

---

[*] See combat report in appendices.

111

Arming a Hawker Tempest V at Volkel.

height, and speed—the standard anti-Mosquito procedure. Without warning, his *Uhu* was suddenly raked by cannon fire. The port engine immediately caught fire. The attacker was a Mosquito N.F. Mk. XXX, MT 281, of No. 410 Squadron RCAF based at Amiens-Glisy in France. The machine was crewed by Flight Lieutenant B.N. Plumer, DFC, and his British navigator, Flight Lieutenant Collins.

Thurow threw the damaged He 219 into a turn and dove away. Plumer followed the He 219, however, and at 5,600 meters he was forced to abandon the pursuit because his protesting Mosquito was exceeding the 800 km/h mark. Thanks to the He 219's diving ability, Thurow was able to elude his pursuer. He extinguished the fire and set course for Handorf on one engine.

When the compressed air system for the ejection seats and then the hydraulics went out, Thurow realized that he would have to make a belly landing. After a successful forced landing Thurow and Neffe

climbed out of their battered but repairable machine. As they did so they saw another Heinkel approaching the airfield. The pilot of the aircraft was attempting a wheels-up landing, but it suddenly lost altitude, crashed, and exploded.

After the death of Oblt. Nabrich, Fritz Habicht flew as radio/radar operator for *Hauptmann* Alexander Count Rességuier de Miremont. On the evening of 3 February over the Ruhr they were pursuing a Lancaster coned by four searchlights. As the attack was about to begin, two of the searchlights suddenly moved and illuminated the He 219. In the ensuing exchange of fire the gunners of the Lancaster set the *Uhu* on fire, while the guns of the German night fighter inflicted fatal damage on the bomber. The *Uhu* may also have been hit by fire from other bombers, however, there is no way to be sure. The four-engined bomber went down near Roermund. The Heinkel, too, went down in a dive. Habicht jettisoned the canopy and Count Rességuier

He 219 A-0 G9 + BA of NJG 1 at Münster-Handorf in the late summer of 1944.

ejected. Habicht then reached down to his right to grasp the ejection handle, but to his horror he realized that the handle had been shot off. His thoughts immediately turned to comrades who had tried to bail out of the He 219 without use of the ejection seat. All had either been torn apart by the propellers or had struck the tail surfaces.

After a desperate struggle, during which the slipstream several times forced him back, Habicht finally got free of the aircraft and immediately pulled the rip-cord. Because of the low altitude at which he had bailed out, Habicht came down in some tall trees almost at the same time as his parachute deployed. While still in the aircraft Habicht had been hit in the left shoulder and chest, but his worst injuries were the result of hitting the trees. Habicht had been extremely lucky. He was one of the few airmen to survive bailing out of the He 219 without use of the ejection seat. This mission ended his war. Habicht's final score was 17 confirmed victories.

On the night of 21 February the British lost 25 heavy bombers in a terror raid on Dortmund. The next night saw Bomber Command launch a double raid on Worms which cost it a total of 62 four-engined bombers. That evening 800 heavy bombers entered Reich airspace, after which 450 turned toward Duisburg. 129 night fighters of Na*chtjagdgeschwader 1, 2, 3, 4* and *6* took off to intercept the raiders. Bitter duels developed over a wide area in the bright moonlit sky. Of the 62 victories claimed by the German defenders, 28, or 45%, went to the credit of just four night fighter pilots. One of these was *Hauptmann* Hager of II./ NJG 1, who brought down eight of the enemy in just 17 minutes. Ground control vectored the night fighters to the bomber stream, and all over the Ruhr Region Bf 110s, Ju 88s, and He 219s clung to the bombers. Every night fighter frequency was extremely active.

Once again the Mosquito long-range fighters posed the greatest threat to the

German night fighters. South of Duisburg Hptm. Schirrmacher and Fw. Waldmann were circling in He 219 *Werknummer* 190211, searching for a target, when they were attacked by a Mosquito. G9 + TH was left badly damaged and unable to maintain altitude. Schirrmacher carried out a forced landing in which Waldmann was injured.

For I./NJG 1 this was the last documented crash by an He 219 in which the crew suffered casualties.

On 21 March 1945 1,000 bombs fell on the night fighter base at Münster/Handorf, and on the following day it was continually strafed for four hours. Seven *Uhus* of I./NJG 1 were destroyed and another 13 seriously damaged. This massive pressure resulted in the transfer of the *Gruppe*. Bremen-Neulandsfeld was supposed to be the unit's next base of operations, but it lacked airfield lighting and was therefore unsuitable for night missions. On 1 April, therefore, *I. Gruppe* moved to Westerland, on the island of Sylt.

On 29 March 1945 there occurred an accident which would not be explained until twenty-five years later. On 26 March 1970 proof was discovered that *III. Gruppe* of NJG 1 had also been equipped with the He 219, at least in small numbers. On that day the Detmold Weapons Disposal Unit recovered an aircraft wreck from a pond which supplied water to the fire-brigade of Isingdorf, near Werther. In the wreckage were found the remains of the bodies of two airmen, along with an identity disc, two parachutes, a leather flight suit, and a well-preserved pistol.

After the second identity disc was finally found it did not prove difficult to identify the two fallen. An investigation revealed the following:

On 29 March 1945 *Unteroffizier* Adam Holl and his radio/radar operator Helmut Walter took off from Dortmund airfield to fly to Hildesheim. Their aircraft was an He 219 A-5, in which they could not

On 21 March 1945 a devastating air raid left the night fighter base at Münster-Handorf severely cratered. This photograph was taken by a Mosquito reconnaissance aircraft one day after the attack.

He 219 A-2 of 1./NJG 1 on 28 November 1944. the painting depicts the aircraft after an attack by Tempests shortly before it crashed at Reckenfeld.

He 219s at Münster-Handorf in 1945, intentionally blown up or wrecked in a bombing raid. In the foreground is G9 + DL of 3./NJG 1.

have had much experience. Why the transfer flight was conducted in broad daylight is not known, but what is certain is that at 1230 hours the night fighter was intercepted by American fighters. The Heinkel had no chance—the machine and its crew were reported missing.

On 9 April 1945 the *Luftwaffe* had available a total of 472 night fighters with which to defend the Reich. Although the early warning system continued to function, most units were grounded by an acute fuel shortage, a situation which was to become even worse.

By this time I./NJG 1 had been reduced to *Staffel* strength. On 9 April it had a strength of 22 He 219s, of which 19 were serviceable. The *Gruppe* was now renamed *1. Staffel*. All of the remaining *Staffeln* were disbanded and their machines taken over by the *Geschwaderstab*.

The *Stab*, based at Eggenbeck, now had a mixed complement of 29 Bf 110s and He 219s, of which 25 were serviceable. At Flensburg, 7./NJG 5, which was made up of the remnants of III./NJG 5, also included a few He 219s in its complement of 32 operational aircraft.

The last victory over an He 219, and a noteworthy one, went to No. 85 Squadron RAF. During the night of 13-14 April 1945 Flight Lieutenant K. Vaugham and his navigator, Flight Sergeant R.D. McKinnon, were flying armed reconnaissance over northern Germany at 9,000 meters in their Mosquito N.F. Mk. XXX. Their Mark X radar detected a fast-moving target two kilometers below them. The Mosquito dove after the unidentified aircraft. A brief burst was sufficient to set the twin-engined machine ablaze. In the glow of the fire Vaugham and McKinnon recognized the aircraft as an He 219 with a turbojet engine mounted beneath its fuselage. Their claim was confirmed by the Senior Intelligence Officer of RAF Station Swannington. The machine may have been the A-010/TL, which Heinkel had been testing for months in a desperate effort to achieve an increase in maximum speed.

The beginning disintegration in the Reich was also affecting Ernst Heinkel AG. RLM correspondence no longer reached the factory. There was little time for new designs such as turbojet-powered night fighters. In April Siegfried Günter, Heinkel's chief designer and its project engineer, moved to Landsberg/Lech. That month the last six He 219s were constructed from spare parts in a forest factory near

Werl. Instead of the DB 603 they were fitted with Jumo 213 engines. The machines' designation was A-7/R5.

On 19 April British troops reached the Elbe near Dannenberg, and on 24 April they linked up with Russian spearheads at Nauen. The Reich capital was thus encircled. The *Luftwaffe* continued to operate from air bases along the Elbe, even though Fighter Command's attacks became heavier by the day. Finally, these bases, too, had to be evacuated.

One unusually active field was Lübeck-Blankensee, which served as a collection point. The blast pens around the airfield were filled with a mixture of Germany's most advanced warplanes: Me 262s, Ar 234 Bs, He 162s, Fw 190 Ds, and Ju 88 night fighters. Everything that could still fly found its way there. Allied fighters were constantly prowling near the airfield, ready to dive on any German aircraft from a safe height and destroy it. Approaches were relatively safe, for the approach lanes and air-

Münster-Handorf, summer 1945. Minus its propellers and nosewheel tire, and its cockpit glazing wrecked, He 219 A-0 W.Nr. 190038, G9 + BH, sits at a dispersal of the former air base waiting to be scrapped.

field were heavily defended by quadruple flak. The anti-aircraft defenses were so good that they posed a danger to any aircraft which the Hungarian flak crews did not recognize. One day an approaching He 219 was fired on, causing its pilot to turn and climb, jettison his canopy, and eject. His parachute did not open until he was 30 meters from the ground. The pilot had been

alone in the aircraft, which dove vertically into a hangar and exploded. Two people were killed on the ground.

On 30 April 1./NJG 1 received orders to disband. All *Staffel* members were to participate in the final battle for Schleswig and Husum as infantry. The war ended before the night fighter personnel could reach Husum.

The End

Of the approximately 274 He 219s manufactured, 54 were found in flyable condition in the British occupation zone in Germany and in Scandinavia. Of these 46 were scrapped on the spot. Five machines were ferried to the Royal Aircraft Establishment in Farnborough. The last three were shipped to the USA for test purposes. Two He 219s fell into Soviet hands and were later handed over, minus radar equipment, to the reborn Czechoslovakian air force. Designated LB-79, the two aircraft were used for experimental missions by the Prague Flight Test Institute until 1952.

Apart from its ejection seats and radar equipment, the He 219 appears to have aroused little interest among the Allies. Captain Eric M. Brown, RN, later Lieutenant Commander and head of the RAE in Farnborough,[*] stated that he never ordered any special tests or evaluation flights in an He 219. He personally flew several *Uhus* from Karup/Grove, Denmark, to the RAE.

Brown kept a notebook in which he recorded his impressions of every aircraft he flew, together with brief notes on especially positive or negative characteristics. His

_____

[*] See account on Page 155.

G9 + SK, an abandoned He 219 A-7 of *2. Staffel*, was still at Handorf in the summer of 1945. Note the bombs lined up in front of the machine.

This He 219 A-5, W.Nr. 420331, of Stab I./NJG 1 was photographed on the Westerland/Sylt airfield after the surrender.

He 219 W.Nr. 310189, G9 + CH, sits on the Westerland/Sylt airfield with many other German aircraft. It was captured in flyable condition by the British.

Westerland/Sylt 1945. He 219 A-2 G9 + TH in night close-support finish.

rating system went from 1 (extraordinary) to 8 (very bad). Brown's notes on the He 219: "He 219 A-2, Note 6, underpowered, poor controllability about the longitudinal axis at low speeds, constant danger of stalling."

So went the Heinkel He 219 into the history of the air war. It is due to the foresight of General "Hap" Arnold that one example of every World War Two aircraft type which the U.S. government could gain possession of was preserved for posterity. Thus it was that of all the He 219s built, a single aircraft, He 219 A-2 *Werknummer* 290202, was preserved. Little is known about how this machine, FE-614, reached the place where it is currently stored. Project Seahorse was an operation which

was supposed to retrieve technically significant aircraft from Germany and deliver them to the USA. A detachment under Colonel Watson scoured occupied Germany for the most advanced aircraft that the former *Luftwaffe* and German aviation industry had to offer. At Karup/Grove, 25 kilometers from Aalborg in Denmark, they found a British-occupied airfield which was intact and being guarded by *Luftwaffe* personnel. In addition to Ar 234s, Ta 152s, and Ju 88 Gs, eight He 219s which German mechanics had maintained in flyable condition were handed over to the British and Americans. Three were assigned the American captured aircraft numbers FE-612 to FE-614. On 26 June 1945, after being briefed by German personnel, two of Wat-

Another photograph of *Werknummer* 310189, now in the hands of the RAE Farnborough as AM 22. The marking VI beneath the cockpit canopy indicated the presence of the FuG 220 Lichtenstein SN-2b with *Streuwelle VI* (dispersal waveband). Another indication of this was the oblique-mounted antennas.

son's pilots, Capt. Fred B. McInstosh and Capt. E.D. Maxfield, took off for Cherbourg. There the British aircraft carrier *Reaper* waited for its priceless cargo. The flight was completed without incident. One day later the third He 219, flown by ferry pilot Hptm. Braun, arrived at Cherbourg airfield. There all sensitive components, such as propellers and antennas, were removed. After receiving a thorough cleaning, each aircraft was sprayed with Cosmoline, a latex coating which would protect the aircraft from corrosion during their sea journey. *HMS Reaper* put to sea on 19 July 1945. On board were 38 German aircraft, plus large quantities of spare parts and engines needed to keep the machines airworthy. *Reaper* reached the port of New York

on 30 July, and the ship's valuable cargo was immediately unloaded. Cranes lifted the three He 219s from the flight deck and placed them on barges for transport to Newark airfield in New Jersey. At the beginning of August McIntosh and Maxfield ferried the three He 219s from Newark to Freeman Field in Seymour, Indiana. One of the machines developed problems with the nose gear hydraulics, which suffered from a buildup of excess pressure. Maxfield solved the problem by threading a wire loop through a hole in the cockpit floor to a valve in the nosewheel bay. From time to time he pulled on the wire to relieve the excess pressure. The aircraft landed at Freeman Field without incident. Pilots 1st Lieutenant James K. Holt and Capt. Raymond

Pilot's position of FE 614, in storage since 1960. These photographs were taken in February 1994 (above left).

Instrument panel and armorglass windscreen of the last He 219 in the world (top center).

Starboard console of the He 219 A-2 (bottom left).

Pilot's cockpit port console. All control levers and instruments bear English stenciling applied after the war (bottom center)

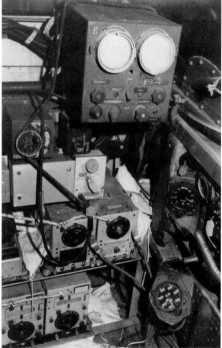

The radar operator's position: the radar equipment appears to be complete, with the exception of the radar scope visors (top right).

Starboard console with fuel gauges, ejection seat handle, and ignition switches (top left).

All equipment appears to be in good condition. Above right the flare pistol firing port.

White flew the He 219 in several test programs for the Foreign Equipment Branch (FEB) until the Air Force's interest in the type waned. FE-613 made its last flight at the beginning of July. Holt ferried the machine to Orchard Place (O'Hare Airport), Chicago. There it was disassembled for storage and later scrapped. While FE-612 was scrapped at Freeman Field, the last intact He 219 stood in the open there until 1951, when it was given to the national Air and Space Museum in Washington, D.C. In 1960 museum workers disassembled FE-614, now NASM 1144, and stored it in a shed of the Paul E. Garber Facility at Silver Hill, Maryland. For long years the aircraft lay quietly beside the famous B-29 "Enola Gay," the world's first nuclear bomber, until the latter was completely restored. The sole surviving He 219 still awaits restoration.

127

FE-614 in 1994

As a branch facility of the Smithsonian Air and Space Museum, the Paul E. Gerber Facility is accessible to the public. Each day visitors view more than 160 aircraft and countless displays of engines and equipment. But not everything is accessible. In spite of repeated requests and reference to work in progress on a book, I was not granted access to the place where the last He 219 is stored. The following series of photos depicting this aircraft in its present state were provided by the late historian Walter Schick. He had the contacts which I lacked on my visit to Silver Hill.

According to the museum the He 219 A-2 is the former *Werknummer* 290202,

which bore the German radio call sign GI + KQ. If this aircraft was not a prototype, and its capture at Karup Grove suggests that it was not, I consider this code questionable at least. G9 + would be more logical, for I./ NJG 1 trained its future aircraft crews in its own Replacement Training *Staffel*. As well, the last letter does not fit with German *Geschwader* coding; internally, this letter was assigned exclusively to the Replacement Training *Staffel*. It is also possible that the letters were reproduced incorrectly, which unfortunately happens all too often in U.S. restorations of German aircraft. The code could have been G9 + QK, in which case everything would be set right again. If the aircraft was an operational ma-

Close-up photograph of the cockpit. Peeled-off layers of paint suggest that tape was once used to seal access panels.

The complete fuselage, with wing and tailplanes stacked beside it. The crosses and swastikas are of American origin, applied before the aircraft was placed in storage.

chine of 2./NJG 1 this would also explain the presence of the complete FuG 220 "Lichtenstein" SN-2b system, including tail warning radar.

Apart from these details the aircraft is in acceptable condition given its age. With the exception of a few details it is complete, and once restored could be a prized part of the collection, unique in all the world.

One of the aircraft's two DB 603 A engines.

The starboard wing and engine nacelle. Clearly visible are the quick-release fasteners developed by Heinkel, which simplified engine installation.

PAGE 132 IS BLANK

Appendices

134 He 219 Loss List

136 Other known *Werknummer*

137 He 219 Loss Table

137 Brief Description of the He 219

138 Specifications of He 219 Variants

146 List of Variants

147 Radar Equipment Carried by the He 219

148 He 219 Cockpit

149 List of *Werknummer*

149 Production Numbers

150 Memories of an He 219 Night Fighter Pilot

155 Test Report by RAE Pilot E. Brown

158 Sources

160 Bibliography

He 219 Loss List

Date	Time	Unit	Crew	Fate	Type	W.Nr.	Code	Remarks
1943								
11-12/6		St./NJG 1	Maj. Streib, Werner	WIA	V9	219009	G9 + FB	Crashed on
			Uffz. Fischer, Helmut	WIA				landing at Venlo
10/7		EHAG	Pilot Könitzer	+	V2	219002	GG + WG	Crashed during
			Ing. Consten					trials, at Mühlleiten
5-6/9	01.45	3./NJG 1	Oblt. Strünning, Heinz	WIA	A-0	219010	G9 + FL	Air combat,
			Obfw. Bieler, Willi	KIA				Jülich
27-28/9	St./NJG 1		Hptm. Frank, Hans Dieter	KIA	A-0	190055	G9 + CB	Collision with
			Obfw. Gotter, Erich					Bf 110, 25 km NW of Celle
20/10	21.11	St./NJG 1	Lt. Schön, Walter	KIA	A-04	190054	G9 + CB	Air combat,
			Uffz. Marotzke, Georg					Klein Ballerstedt/Stendal
1944								
21-22/1	23.00	St./NJG 1	Hptm. Meurer, Manfred	KIA	A-0	190070	G9 + BB	Air combat, 20 km
			Obfw. Scheibe, Gerhard					E of Magdeburg
28/3		E-Stelle	Lt. Jung		A-0	190052		Crashed, Treptow
7/5	07.05	1./NJG 1	Fw. Heinzelmann, Emil	KIA	A-0	190115	G9 + FH	Ferry flight,
			Uffz. Herling, Wilhelm	KIA				crashed at Süchteln
21/5	17.00	1./NJG 1	Uffz. Tampke, Ewald	WIA	A-0	190107	G9 + FL	Air combat, 20 km S
			Uffz. Tanbs, Eduart	WIA				of Herning
1/6	14.00	3./NJG 1	Oblt. Guth, Fritz	KIA	A-0	190119	G9 + AK	Crashed 2 km E
			Fw. Klein, Andreas	KIA				of Muldbjerg
			Obgefr. Otto, Herbert	KIA				
3-4/6		3./NJG 1	Hptm. Eicke, Heinz	A-0	190188		G9 + BL	Air combat, N of
			Obfw. Gall, Heinz	KIA				Wilhelminsdorp
5-6/6	00.59	3./NJG 1	Lt. Mauß, Ernst	WIA	A-0	190177	G9 + IK	Engine failure,
			Uffz. Kraus, Günther	+				crashed 8 km E of Herning
15/6			Obfw. Blöde	A-093	190022			Crashed, Jürgenau
15-16/6	02.15		Uffz. Beyer, Willi	+	A-0	190180	G9 + RK	Crashed 1 km N
			Obgefr. Walter, Horst	+				of Leersum
1/7		St./NJG 1	Not known	A-0	?		G9 + CB	Crashed
14/7	15.45	2./NJG 1	Oblt. Finke, Joachim	+	A-0	190212	G9 + LK	Crashed Venlo
			Uffz. Petzold, Felix	+				
22/7		3./NJG 1	Hptm. Eicke, Heinz	WIA	A-0	190185	G9 + KL	Engine failure,
			Fw. Feliner-Feldegg, Hugo					crashed Schleswig
9/9	17.55	2./NJG 1	Obfw. Jadatz, Heinz	KIA	A-2	290105	G9 + DK	Air combat,
			Uffz. Schindler, Alfred	KIA				Hopsten
			Uffz. Wennholz, Heinrich	KIA				
9/9	17.58	2./NJG 1	Fw. Wildhagen, Karl	KIA	A-0	190128	G9 + OK	Air combat,
			Uffz. Kramer, Gustav	WIA				crash-landing Hopsten
			Obgefr. Ociepka, Wilhelm	WIA				
			Gefr. Neumeier, Heinz	WIA				
25-26/9		2./NJG 1	Lt. Schirmer, Günther	WIA	A-0	190193	G9 + EK	Crash-landing at
			Gefr. Rosenberger, Wilhelm	WIA				Münster-Handorf
1/10		St./NJG 1	Maj. Förster, Paul	+	A-0	190194	G9 + CL	Crash-landing at
			Oblt. Apel, Fritz	+				Handorf
14-15/10	20.00	1./NJG 1	Uffz. Frankenhauser, Franz	A-0	190059		G9 + EH	Bailed out after
			Uffz. Biank, Helmut	KIA				air combat, Münster
15-16/10		1./NJG 1	Oblt. Stieghorst, Paul	KIA	A-2	290002	G9 + BH	Air combat, 8 km
			Uffz. Frunske, Kurt	KIA				NE of Bremen
/11			Pilot Lehnhoff	A-0	190234			Practice flight, Welzow
4-5/11	20.00	3./NJG 1	Obfw. Morlock, Wilhelm	KIA	A-0	190182	G9 + HL	Air combat,
			Fw. Soika, Alfred	WIA				Ibbenbüren
28/11	08.35	2./NJG 1	Lt. Fischer, Kurt Heinz	KIA	A-2	290010	G9 + MK	Shot down at
			Uffz. Bauer, Hermann	KIA				Greven/Reckenfeld during
								practice flight
30/11-1/12		1./NJG 1	Uffz. Frankenhauser, Franz	+	A-2	290061	G9 + CH	Engine failure, crashed
			Uffz. Fabian, Erwin	+				3 km E of Sendenhorst
4/12		2./NJG 1	Obfw. Luchtmayer, Manfred	+	A-5	420322	G9 + NK	Crashed at
			Uffz. Posselt, Heinz	+				Ostbevern
5-6/12	22.05	2./NJG 1	Uffz. Wollenhaupt, Werner	+	A-2	290129	G9 + CK	Crashed at
			Uffz. Heimesaat, Günther	+				Telgte
14-15/12	22.47	1./NJG 1	Oblt. Tonn, Wolfgang	+	A-0	190234	G9 + NH	Crashed on approach,
			Obgefr. Wolff, Joachim	+				12 km N of Handorf

Date	Time	Unit	Crew	Fate	Type	W.Nr.	Code	Remarks
17/12	18.00	NJG 1			A		-G9 +	Crashed into building at Handorf while landing after air combat
17-18/12		3./NJG 1	Lt. Prietze, Jürgen	KIA	A-2	290188	G9 + WL	Air combat, 2 km
			Uffz. Haake, Frithjof	KIA				SE of Sonsbeck
18-19/12	22.15	1./NJG 1	Uffz. Scheuerlein	WIA	A-0	190229	G9 + GH	Air combat,
			Uffz. Heinze, Max Günther	WIA				Südlohn

1945

Date	Time	Unit	Crew	Fate	Type	W.Nr.	Code	Remarks
31/12-1/1		3./NJG 1	Oblt. Oloff, Heinz	WIA	A-2	290194	G9 + KL	Air combat
			Fw. Fischer, Helmut	WIA				Schleiden, crew bailed out
5-6/1	20.08	2./NJG 1	Obfw. Ströhlein, Josef	KIA	A-0	190188	G9 + CK	Air combat, 5 km
			Uffz. Kenne, Hans					S of Wesendorf
5-6/1		1./NJG 1	Obfw. Keilich, Hans	WIA	A-5	420329	G9 + BH	Engine failure, crashed at Diepholz
7-8/1		1./NJG 1	Uffz. Barkenfeld, Karl	+	A-2	290072	G9 + FH	Crashed at Bad
			Obgefr. Freudenberger, Ernst	+				Iburg
14/1					A-			Air combat, 8 km E of Hamm
14/1	09.28	3./NJG 1	Lt. Lehr, Reinhold	KIA	A-2	290125	G9 + PL	Air combat, 9 km
			Obgefr. Leukam, Albert	KIA				N of Münster
3-4/1	20.30	1./NJG 1	Hptm. Graf Ressugier, Alex		A-2	290070	G9 + CH	Air combat with Lancaster,
			Fw. Habicht, Fritz	WIA				Roermund, crew bailed out
27/1		NJG 1			A-5	420320	G9 +	Strafing attack, Ludwigslust
1/2		NJG 1	Not known		A-0	190210	G9 +	Crashed Bad Honnef
1/2		II./NJG 1	Not known		A-2	210004	G9 +	Crashed Paderborn
3/2		NJG 1	Not known		A-0	190235	G9 +	Crashed Handorf
3/2		1./NJG 1			A-2	290004	G9 + DH	Strafed at Handorf
3/2		NJG 1	Not known		A-2	290058	G9 +	Crashed near Handorf
3/2		NJG 1			A-7	310190	G9 +	Bombing raid, Ludwigslust
3/2		NJG 1			A-7	310202	G9 +	Bombing raid, Ludwigslust
3/2		NJG 1			A-7	310214	G9 +	Bombing raid, Ludwigslust
6/2		NJG 1	Not known		A-5	420324	G9 +	Crashed near Handorf
7/2		NJG 1	Not known		A-2	290203	G9 +	Crashed near Telgte
21-22/2		1./NJG 1	Hptm. Schirrmacher		A-0	190211	G9 + TH	Air combat
			Fw. Waldmann, Franz	WIA				
29/3	12.30	8./NJG 1	Uffz. Holl, Adam	KIA	A-5	420...	G9 +	Shot down during
			Uffz. Walter, Helmut	KIA				ferry flight, Isingdorf
?		not known	Pilot Susemühl		A-7	310322	?	Not known
May 45		EHAG			A-0	190062	RL + AB	Scrapped, Vienna
May 45		E-Stelle			A-0	190113	DV + DI	Scrapped, Riem (A-019)
May 45			not known		A-0	190176		Captured by Americans at Lechfeld
May 45		1./NJG 1			A-2	290123	G9 + TH	Captured by British at Westerland/Sylt
May 45		3./NJG 1			A-7	310189	G9 + CL	Captured by British at Westerland/Sylt
May 45		St./NJG 1			A-5	420331	G9 + DB	Captured by British at Westerland/Sylt
May 45		NJG 1			A-7	310109	G9 + VH	Captured by British at Westerland/Sylt
May 45		NJG 1			A-7	310215	G9 +	Captured by British at Westerland/Sylt
May 45		NJG 1			A-2	290062	G9 +	Captured by British
May 45		NJG 1			A-7	310126	G9 +	Captured by British
May 45		NJG 1			A-7	310106	G9 +	Captured by British
May 45		1./NJG 1			A-2	290004	G9 + DH	Blown up at Handorf
June 45		NJ.Erg.			A-2	310350	?	Captured by Americans at Grove
June 45		NJ.Erg.			A-2	290060	?	Captured by Americans at Grove
June 45		NJ.Erg.			A-2	290202	GI + KQ	Captured by Americans at Grove

Other Known Werknummern or Signs

Date	Unit	Crew	Type	W.Nr.	Code	Loss/Remarks
6/11/42	EHF	Pilot Petersen	V1	219001	VG + LW	Maiden flight
10/7/43	EHF	Pilot Könitzer	V2	219002	GG + WG	Diving tests, written off in crash
19/4/43	EHF	Pilot Schuck	V3	219003	?	Crash-landing during trials
43-44	E-Stelle		V4	190004	DH + PT	Power plant tests
43-44	E-Stelle		V5	190005	GE + FN	MK 108, Jumo 222 and 4-blade propellers
43-44	E-Stelle	Pilot Eisermann	V6	190006	DH + PV	Ejection seat tests
43-44	NJG 1		V7	190007	G9 + DB	Service trials Venlo
43-44	NJG 1		V8	190008	G9 + EB	Service trials Venlo
11/6/43	NJG 1	Pilot Streib	V9	190009	G9 + FB	Service trials Venlo, written off in crash
43-44	NJG 1		V10	190010	G9 +	Service trials Venlo
43-44	EHF		V11	190011	?	Diving tests
43-44	NJG 1		V12	190012	G9 +	Service trials Venlo
43-44	E-Stelle		V13	190052	?	Performance measurements
43-44	E-Stelle		V-14	190058	?	Performance enhancement, tail section experiments
14/5/44	E-Stelle	Pilot Eisermann	V15	190064	RL + AD	Tests with GM1, FuG 16ZY
43-44	EHF		V16	190016	RL + AJ	Tests with Jumo 222, larger wing
43-44	EHF		V17	190060	PK + QJ	Service trials DB 603 with G-supercharger, crash-landing Vienna
43-44	EHF		V18	190071	BF + JF	Jumo 222 and 4-blade propeller
4/44	EHAG		V19	190073	?	Planned 3-man cockpit, cancelled
21/4/44						
43-44	EHF		V20	190020	?	Experiments with pressurized cockpit
43-44	E-Stelle		V25	190122	?	Single-core cables
43-44	NJG 1		V26	190120	G9 +	MK 108 oblique armament
43-44	E-Stelle		V28	190068	VO + BC	Tactical brake (braking chute)
43-44	E-Stelle		V29	190069	?	Power plant test-bed
43-45	E-Stelle	Pilot Cap Ing. Schalda	V30	190101	?	A-010 with BMW 003 jet engine. 40% damage after crash-landing. Machine may have been written off when shot down on 13-14/4/45.
15/5/44	EHAG		V31	190106	DB + DT	Load measurements, special loading
43-44	E-Stelle		V31	190106	DV + DB	Tactical brake (braking chute)
43-44	NJG 1		V32	190121	G9 +	GM 1 system
43-44	E-Stelle (Telefunken)		V33	190063	DV + DL	Antenna test-bed
5/44	EHAG		V34	190112	?	3-man cockpit, increased range
17/5/44	EHAG		V76		BE + JC	Autopilot test flight
43-44	E-Stelle	Pilot Eisermann	A-0	190113	DV + DI	Ejection seat trials
44	E-Stelle		A-0	190062	RL + AB	Flame damper tests
44-45	E-Stelle		A-0		CS + QI	Me P 8 propeller tests
44	EHF		A-0	190068	RL + AH	V28, converted with shortened wingspan
5/44	St./NJG 1	Hptm. Förster, Paul	A-0	190004	G9 + AB	Venlo
4/44	2./NJG 1	Oblt. Modrow	A-0	1900..	G9 + FK	Venlo
6/44	1./NJG 1	not known	A-2	290068	G9 + SH	Ferried from Vienna to Venlo
44	E.Kdo.Lärz		A-0	190051	?	Long-term testing
44	E.Kdo.Lärz		A-0	190055	?	Long-term testing
44	E.Kdo.Lärz		A-0	190057	?	Long-term testing
44	E-Stelle		A-0	190059	?	Generator experiments
44	E-Stelle		A-0	190061	?	De-icing and cockpit heating tests

Table of He 219 Losses

| | Operational losses over Germany | | | Other losses over Germany | | | | | | |
| | Flak | | | Fighters | | | Bombing* | Various causes | | |
	A	B	C	A	B	C	A	A	B	C
06/1944	—	—	2	1	—	—	—	4	3	3
07/1944	—	—	—	1	—	—	—	2	7	5
08/1944	—	—	1	1	—	—	—	1	7	3
09/1944	—	—	—	1	—	3	1	1	2	1
10/1944	—	—	—	1	—	—	—	3	1	2
11/1944	—	—	—	1	—	—	—	1	1	3
12/1944	—	—	—	5	—	—	—	6	—	2
01/1945	—	—	—	7	—	1	1	4	—	6
02/1945	1	—	1	5	2	1	—	1	4	2
03/1945	1	—	1	—	—	—	—	—	—	—
04/1945	no information									
05/1945	no information									

Legend: A; Total loss B: Damaged** (replacement parts required) C: Lightly damaged (no replacement parts required)
* Only aircraft destroyed in bombing raids are listed, He 219s which were destroyed on factory airfields or in factories before they could be handed over to the *Luftwaffe* are not included.
** Aircraft in category B which were too badly damaged to be repaired in the workshop were sent to the aircraft repair facility of the Flugzeugwerke Eger GmbH in the Sudetenland (usually by rail).

Note: All He 219s which were taken to England were scrapped after extensive testing. The following dates are known:

Date	Unit	Type	W.Nr.	Code	Loss/Remarks
21/08/45	1./NJG 1	A-2	290123	G9 + TH	Scrapped Brize Norton
30/08/45		A-7	310109	G9 + VH	Scrapped Brize Norton
../10/45		A-7	310106		Scrapped Brize Norton
../01/48	3./NJG 1	A-7	310189	G9 + CL	Scrapped Brize Norton
../12/48		A-7	310215		Scrapped Brize Norton

Brief description of the He 219

A. General

Twin-engined, all-metal, cantilever mid-wing monoplane. Forward projecting nose section, all-round-vision canopy, tapering rear fuselage, tall twin fins and rudders, trapezoid-shaped wing with slender profile, conventional engine arrangement forward of wing, rearwards-retracting main undercarriage elements with twin wheels and a rearwards-retracting nosewheel.

B. Significant Design Features
Fuselage

All metal (monocoque) construction. Rectangular cross-section with rounded corners (tapering to the rear). Nose section (cockpit) bolted to fuselage center section. Fuselage components riveted. Three fuel tanks in fuselage center section. Cockpit and tanks separated by bulkhead rib (fuselage walls parallel in this area). Cockpit partially armored.

Wing

Cantilever, one-piece wing with main and end spars and removable wingtips. The engine nacelles formed part of the wing, their stressed-skin shells also serving as engine bearers. Undercarriage attachment points in symmetrical arrangement in center of engine nacelles on main and end spars. Split flaps and slotted ailerons arranged on trailing edge of entire wing.

Control Surfaces

The horizontal stabilizer is a cantilever, one-piece structure. The stabilizer profile is unsymmetrical. The vertical tail consists of vertical stabilizers attached to the ends of the horizontal stabilizer. Ailerons and landing flaps are in the form of slotted ailerons and split flaps. Elevators and rudders are equipped with trim tabs which can be set from the cockpit.

Undercarriage

Two main undercarriage units each with twin wheels, hydraulically retractable rearwards, and a nosewheel which swivels through 90 degrees and lies beneath the cockpit floor when retracted.

Power Plants

Two DB 603 A twelve cylinder liquid-cooled fuel-injected engines with a reduction ratio of 1.93 : 1. Single-stage hydraulic supercharger and automatic boost control. Each engine powers a VDM three-blade metal variable-pitch propeller.

Specifications of He 219 Variants (Factory Figures)

He 219 Prototype **(December 1942)**

Purpose:	night fighter		
Crew:	Two (seated back to back). Access to cockpit by way of a folding ladder which retracts into the left side of the fuselage.		
Dimensions:	Wingspan:	18 500 mm	
	Aerodynamic surface:	44.5 m	
	Length:	15 540 mm	
	Height:	4 400 mm	
	Main undercarriage wheel base:	500 mm	
	Main undercarriage wheel:	840 x 300	
	Nosewheel:	770 x 270	
Loadings:	Wing loading at weight of 11.75 tons:	264 kg/m	
	Power loading at weight of 11.75 tons:	3.36 kg/h.p.	
Power Plants:	Two Daimler Benz DB 603 A, 12-cylinder inverted-vee liquid-cooled engines each producing 1,750 h.p. and driving three-blade VDM propellers (D = 3.6 m).		
Fuel Capacity:	Three fuselage tanks: 1 100 l forward, 500 l center and 990 l aft. The tanks are located directly behind the crew compartment. Fuel: B4 (87 octane)		
	Two lubricating oil tanks located in engine nacelles, each 85 l.		
Weights:	Empty weight:	9 030 kg	
	Crew:	200 kg	
	Fuel:	1 930 kg	
	Lubricants:	142 kg	
	Ammunition:	484 kg	
	Takeoff weight:	11 750 kg	
Performance:	Maximum speed	3,150 h.p. at 0 m	490 km/h
	at combat power	3,000 h.p. at 6 500 m	615 km/h
	Cruising speed*	2,750 h.p. at 0 m	470 km/h
	Time to climb at takeoff weight:	to 2 000 m	3.5 min.
		to 4 000 m	7.2 min.
		to 6 000 m	11.5 min.
		to 8 000 m	18.0 min.
	Service ceiling	9 900 m	
	1/2 fuel	10 850 m	
	Takeoff roll/takeoff distance:	520/780 m	
	Rate of climb after takeoff:	3.9 m/sec	
	Landing speed at weight of 10.4 tons:	150 km/h	
Armament:	6 x MG 151/15		
	Dorsal position, 1 hand-held MG 131		

(* Takes into account fuel consumed in run-up, takeoff, climb and 2% reserve)

He 219 A-0   (Jan. 44) Differences from prototype.

Weights:	Empty weight:	9 570 kg
	Crew, fuel, lubricants, ammunition:	2 690 kg
	Takeoff weight:	12 260 kg

Performance:	Maximum speed:	at 0 m	500 km/h
	At maximum boost altitude:	6410 m	605 km/h
	Cruising speed*	0m	455 km/h
	Service ceiling	9 300 m	

Armament: 2 x 20-mm Mauser MG 151/20 A (300 rounds per gun, one in each wing root.
Equipment Sets: **M1:** 4 x MG 151/20 A in ventral tray each with 300 round per gun
(*Rüstsätze*)   **M2:** 4 x MK 108 30-mm in ventral tray with 300 rounds per gun
                 M3: 4 x MK 103 30-mm in ventral tray with 300 rounds per gun

He 219 A-2   (June 44) Differences from prototype.
Initial production version, oblique armament, additional fuel, flame dampers and radar antennas

Weights:	Empty weight:	9 440 kg	
	Crew, fuel, lubricants, ammunition:	3 060 kg	
	Takeoff weight:	12 500 kg	
Performance:	Maximum speed	at 0 m	460 km/h
	At maximum boost altitude	6 000m	560 km/h
	Cruising speed*	0 m	540 km/h
	Service ceiling	8 900 m	

Armament: 2 x 20-mm Mauser Mg 151/20 A (300 rounds per gun), one in each wing root.
Equipment Sets: **M1:** 4 x MG 151/20 A in ventral tray each with 300 round per gun
(*Rüstsätze*)   **M2:** 4 x MK 108 30-mm in ventral tray with 300 rounds per gun
                 M3: 2 x MK 108 30-mm and 2 x MG 151/20 in ventral tray each with 300 rounds per gun
                 Oblique armament 2 x MK 108 in rear fuselage

He 219 A-5 (Jan. 44) Differences from A-2
Three-man fighter

Power plants: Two Daimler Benz DB 603 E, each 1,800 h.p.

Weights:	Empty weight:	9 665 kg	
	Crew, fuel, lubricants, ammunition:	3 160 kg	
	Takeoff weight:	13 575 kg	
Performance:	Maximum speed	at 0 m	450 km/h
	at maximum boost altitude	6 000 m	585 km/h
	Cruising speed*	0 m	555 km/h
	Service ceiling	9 000 m	

Armament: 2 x 20-mm Mauser MG 151/20 A (300 rounds per gun), one in each wing root.
                 Possible defensive armament of one MG 131
                 Oblique armament of 2 x MK 108 in rear fuselage
Equipment Sets: **M1:** 4 x MG 151/20 A in ventral tray each with 300 round per gun
(*Rüstsätze*)   **M2:** 4 x MK 108 30-mm in ventral tray with 300 rounds per gun
                 M3: 2 x MK 108 30-mm and 2 x MG 151/20 in ventral tray each with 300 rounds per gun

He 219 A-6 (Nov. 44) Differences from A-2
Mosquito-hunter, lightened A-2 with power plant and ammunition canister armor removed, no oblique weapons, flame dampers and radar antennas removed.

Weights:	Empty weight:	8 470 kg	
	Crew, fuel, lubricants, ammunition:	3 060 kg	
	Takeoff weight:	11 800 kg	
Performance:	Maximum speed	at 0 m	475 km/h
	at maximum boost altitude	8 500 m	600 km/h
	Cruising speed*	0 m	595 km/h
	Service ceiling	11 400 m	

Armament: 2 x 20-mm Mauser MG 151/20 A (200 rounds per gun), one in each wing root.
2 x 20-mm MG 151/20 in ventral tray with 300 rounds per gun

He 219 A-7 (Jan. 45) Differences from A-2
Development of A-2 with radar antennas and flame dampers

Power plants: Two Daimler Benz DB 603 G, each 1,900 h.p.
Replaced by: Two Jumo 213 E, each 1,740 h.p.

Performance:	Maximum speed	at 0 m	460 km/h
	at maximum boost altitude	6 000 m	580 km/h
	Cruising speed*	0 m	550 km/h
	Service ceiling	9 000 m	

Armament: 2 x 20-mm Mauser MG 151/20 A (300 rounds per gun), one in each wing root
                 Oblique armament 2 x MK 108 in rear fuselage
                 2 x MG 151/20 A in ventral tray with 300 rounds per gun
                 Increase to 4 x MG 151/20 in ventral tray planned

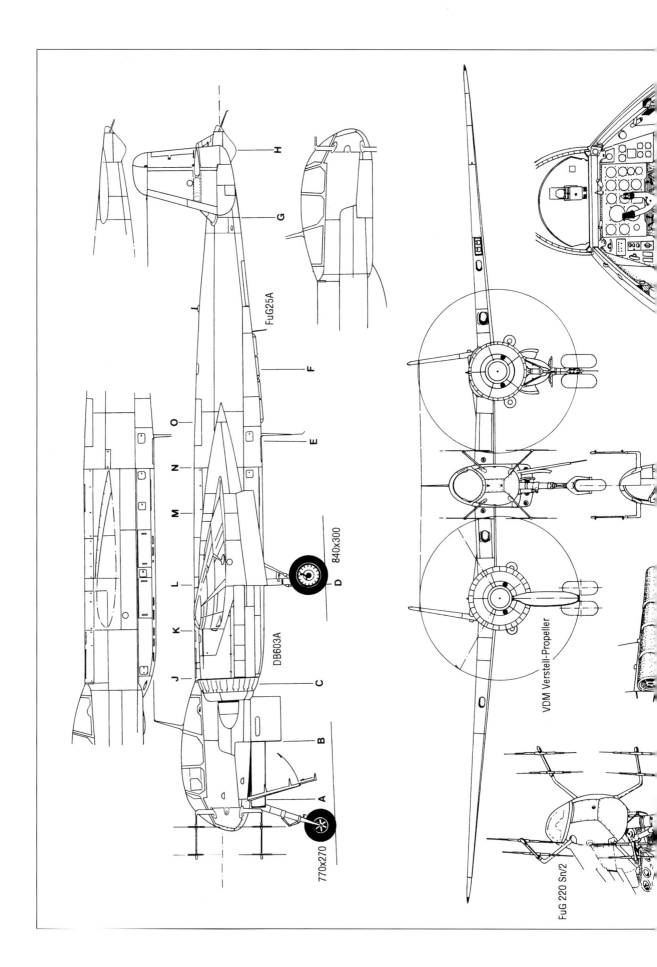

FuG25A

DB603A

770x270

840x300

VDM Verstell-Propeller

FuG 220 Sn/2

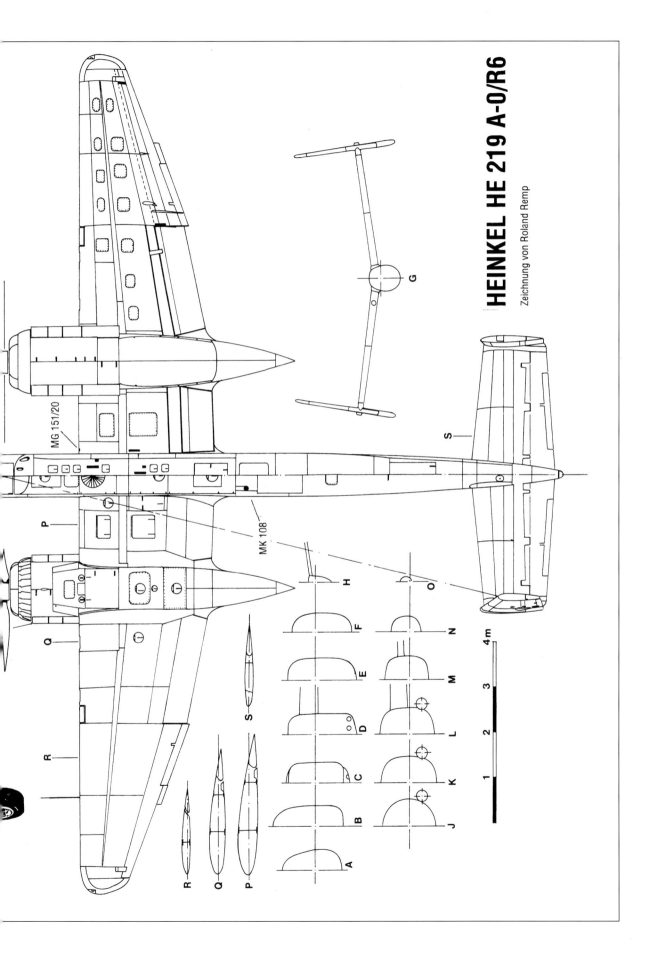

HEINKEL HE 219 A-0/R6

Zeichnung von Roland Remp

MG 151/20

MK 108

Heinkel He 219 A-5/R3

1 **Fresh air inlet for cockpit heating**
2 FuG 220 Lichtenstein SN-2 antenna
3 Armored nose
4 Windscreen
5 Windscreen wiper/washer
6 Handhold
7 Inner armor glass windscreen
8 Revi 18 B gunsight
9 Armored visor (deleted on later models)
10 Control column
11 Revi 16 A-N overhead gunsight for *schräge Musik*
12 Folding headrest
13 Pilot's ejection seat
14 Port instrument console
15 Footstep
16 Entry ladder (extended)
17 Nosewheel leg
18 Nosewheel doors
19 Compressed air bottles
20 Nosewheel retraction cylinder
21 Ejection seat mounting
22 Radar operator's ejection seat
23 Flare pistol port
24 Hinged headrest
25 Antenna mast
26 FuG 212 radar scope
27 FuG 220 radar scope
28 Fuselage frame 9 and 9a
29 Port wing root cannon port
30 Forward fuselage fuel tank (1 100 liters)
31 Fuel filler cap
32 Tank cover with suppressed D/F aerial
33 Main spar connection joint
34 Flame damper tube
35 Liquid coolant tank
36 Propeller drive shaft
37 Propeller hub
38 VDM constant-speed propeller
39 Daimler Benz DB 603 E engine
40 Supercharger
41 Oil tank
42 Propeller de-icing tank
43 Main wing spar
44 Starboard wing heating unit
45 Air intake
46 FuG 101 radio altimeter
47 Starboard navigation light (green)
48 Starboard aileron
49 Wing structure
50 Aileron tab
51 Landing flap
52 Flap actuator
53 Underwing inspection panels
54 Nacelle fuel tank (390 liters)
55 Main undercarriage (retracted)
56 Inboard flap section
57 Mainwheel doors
58 Undercarriage pivot point
59 Firewall
60 Starter fuel tank
61 Center fuel tank (500 liters)
62 Fuel filler cap
63 Fuselage frame 17
64 Wing aft attachment point
65 Port MG 151/20
66 Wing main attachment point
67 Ammunition tank and feed for forward port fuselage cannon
68 Ammunition tank and feed for aft port fuselage cannon
69 Port propeller de-icing tank
70 Oil tank
71 Engine accessories
72 Engine bearer
73 Daimler Benz DB 603 E engine
74 Coolant tank
75 Controllable radiator gills
76 Propeller hub
77 VDM variable-pitch propeller
78 Armored annular radiator
79 Flame damper tube
80 Supercharger air intake

81 Port wing heating unit
82 Flap actuator
83 Aileron control quadrant
84 Landing light
85 Aileron control rod
86 Pitot tube
87 Main wing spar
88 Wing skinning
89 Port navigation light (red)
90 Starboard aileron
91 Fixed trim tab (port side only)
92 Auxiliary aileron tab
93 Main undercarriage
94 Mainwheel doors
95 Undercarriage leg
96 Starter fuel tank
97 Undercarriage retraction jack
98 Pressure-oil tank (port nacelle only)
99 Nacelle fuel tank (390 liters)

100 Starboard undercarriage
101 Rear fuel tank (990 liters)
102 Fuel filler cap
103 Fuselage frame 20
104 Ammunition feed
105 Ammunition tanks (100 rounds)
106 Twin oblique-mounted MK 108 cannon (*schräge Musik*)
107 Electrical supply cables (starboard fuselage wall)
108 Oxygen bottles
109 Maintenance platform
110 Ventral antenna
111 FuG 25A antenna
112 Entry hatch
113 Walkway
114 Electrical navigation equipment
115 Crew escape dinghy
116 D/F loop antenna

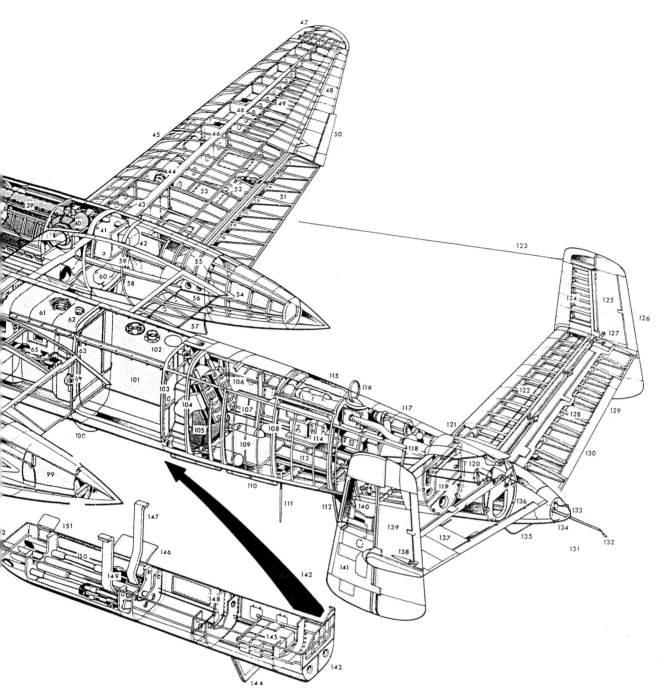

117 BLO 30/U fuselage heating and tailplane de-icing unit
118 Heating ducts
119 Fuselage frame 31
120 Tail unit control linkage
121 Fresh air inlet
122 Tailplane construction
123 Aerial
124 Tailplane construction
125 Starboard rudder
126 Rudder tab
127 Rudder control hinge
128 Elevator construction
129 Elevator trim tab
130 Flettner auxiliary tab
131 Trailing aerial tube
132 Trailing aerial
133 Tail navigation light
134 Tail cone

135 Tail bumper
136 Fuselage frame 33
137 Port elevator
138 Rudder tab hinge fairing
139 Port rudder
140 Tailplane construction (not drawn)
141 Skinning
142 Ventral weapons tray
143 Fuselage frame 20
144 Ventral maintenance hatch
145 Main junction boxes
146 Weapons access hatches
147 Ammunition feeds
148 Mounting space for additional cannon
149 Forward (outboard) MK 108
150 Blast tubes
151 Gun sighting/correction hatch
152 Cannon ports

He 219 A-2/R2

1 Cockpit fresh air inlet
2 Windscreen washer
3 FuG 220 Lichtenstein antennas
4 Revi 16B gunsight
5 Internal armor glass panel with wiper and warm air vent
6 Overhead Revi 16 A-N for oblique-mounted cannon (*schräge Musik*)
7 Jettisonable cockpit canopy (foldable to starboard)
8 Jettisonable radar operator's canopy
9 Pilot's ejection seat
10 Radar operator's ejection seat
11 Compressed air bottles for ejection seats
12 FuG 220 Lichtenstein radar scope
13 FuG 10 P voice radio set
14 Flare pistol
15 Fuselage frame 9/9a
16 First-aid kit
17 Firing handle for radar operator's ejection seat
18 Forward fuel tank 1 100 liters
19 Center fuel tank 500 liters
20 Rear fuel tank 990 liters
21 Fuel lines
22 Fuel filler caps
23 D/F auxiliary antenna
24 Main spar attachment point—fairing.
25 Forward part of ventral weapons tray
26 Port MK 103 in ventral weapons tray
27 Ammunition feed
28 MK 108 in port wing root
29 Forward armor plate for ammunition supply
30 Main spar
31 Ammunition feed
32 Release for wing root MG/MK
33 Twin oblique-mounted 30-mm MK 108 *schräge Musik*
34 Ammunition tanks (2 x 100 rounds)
35 Ammunition feed
36 Compressed air bottles for *schräge Musik*
37 Control rods
38 Electrical supply cables (starboard fuselage wall)
39 Electrical main navigation equipment (altimeter, transformer, master compass, FuB1 2F, FuG 16 ZY, FuG 25 A, FuG 217 Neptun)
40 Ditching equipment with dinghy
41 Loop antenna FuG 16 ZY approach aid
42 Antenna mast FuG 25 A IFF
43 Maintenance hatch for oblique-mounted weapons and electrical equipment
44 Fresh air intake for fuselage heater
45 Fuselage heating and tailplane de-icing unit
46 Tailplane control rods
47 Tail cone
48 Trailing aerial with tube and aerial
49 Navigation light
50 Horizontal stabilizer and elevator
51 Flettner tab
52 Trim tab
53 Rudder
54 Fixed aerial in port tailplane
55 Antenna mast
56 VDM variable-pitch propeller
57 Armored-front annular radiator
58 Controllable radiator gills
59 DB 603 A engine, 1,750 h.p.
60 Coolant tank, 24 liters
61 Flame damper tube
62 Supercharger air intake
63 Supercharger
64 Firewall
65 Oil tank, 85 liters
66 Propeller de-icing fluid tank
67 Pressure-oil tank (port wing only)
68 Main undercarriage
69 Fresh air intake for port wing heater (de-icing)
70 Landing light
71 Aileron control rod
72 Aileron
73 Auxiliary aileron tab
74 Trim tab (port aileron only)
75 Landing flap (lowered)

76 Electrically-heated pitot tube
77 Navigation light (red)
78 Nosewheel
79 Nosewheel door
80 Entry ladder, extended

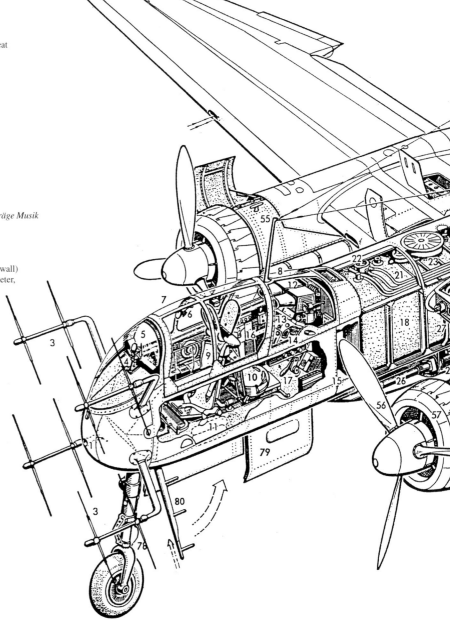

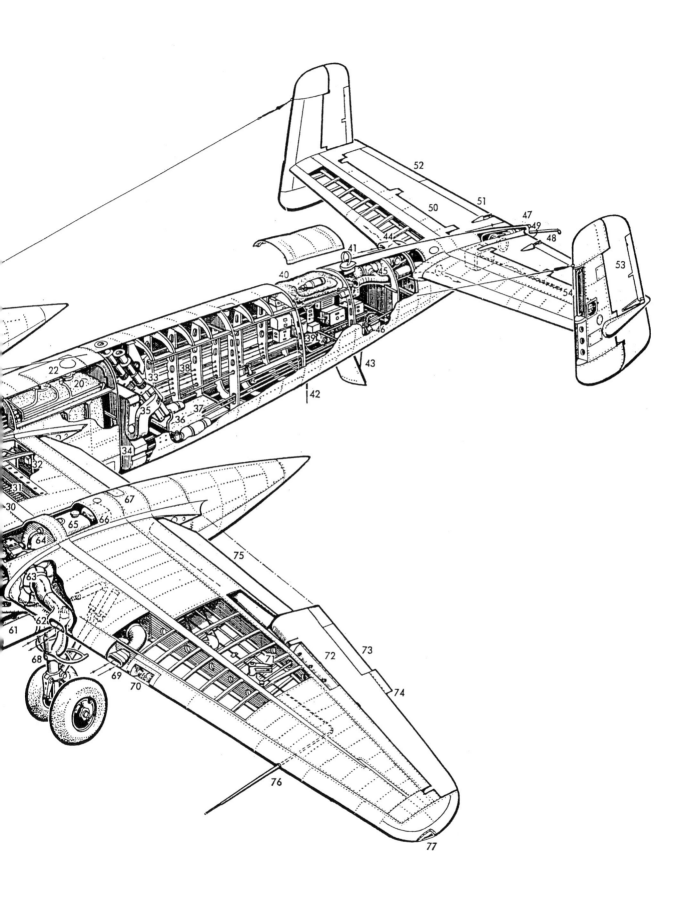

He 219 List of Variants

Type	Description	Power plants	Armament	Radio equipment
A-0/R1	lengthened fuselage, standard pre-production wing, 2-man cockpit	DB 603 A	2 x MK 108 in ventral tray 2 x MG 151/20 in wing roots Weapons sets M1 – M3	FuG 212 C1
A-0/R2	as A-0/R1 strengthened undercarriage	DB 603 A	2 x MK 103 in ventral tray 2 x MG 151/20 in wing roots	FuG 212 C1
A-0/R3	improved A-0 prototype for A-2	DB 603 A	4 x MK 103 in ventral tray 2 x MG 151/20 in wing roots	FuG 212 C2
A-0/R6	prototype for A-5	DB 603 A	4 x MK 103 in ventral tray 2 x MG 151/20 in wing roots	FuG 212 and 220
A-1	planned A-0 production version flattened cockpit canopy	DB 603 A/B	armament sets M1 – M3	FuG 220
A-2/R1	improved A-0, single-core wiring, greater range	DB 603 A/B	4 x MK 103 in ventral tray	FuG 220
A-2/R2	as A-2/R1, 900-l fuel in external tanks, planned as bomber	DB 603 A to DB 603 G	2 x MK 103 in ventral tray 2 x MK 108 in wing roots 2 x MK 108 in *schräge Musik* planned	FuG 220
A-3	development of A-0/R6 series cancelled	DB 603 A/B	2 x MK 108 in ventral tray 2 x MG 151/20 in wing roots	
A-4	development of A-2 for anti-Mosquito, reconnaissance reduced armor and armament, GM 1 system	DB 603 A/B Jumo 222	2 x MK 103 in ventral tray 2 x MK 108 in wing roots	FuG 220
A-5/R1	further development of A-3	DB 603 A	2 x MK 108 in ventral tray 2 x MG 151/20 in wing roots 2 x MK 108 *schräge Musik*	FuG 212 and 220
A-5/R2	prototype for A-7/R4	DB 603 A	2 x MG 151 in ventral tray 2 x MG 151/20 in wing roots 2 x MK 108 *schräge Musik*	FuG 220
A-5/R3	production version	DB 603 E	2 x MK 103 in ventral tray 2 x MG 151/20 in wing roots 2 x MK 108 *schräge Musik*	FuG 220
A-5/R4	conversion of A-5/R1 with 3-man cockpit, increased range, planned defensive armament based on V34	DB 603 E	2 x MG 151/20 in ventral tray 2 x MG 151/20 in wing roots 2 x MK 108 *schräge Musik* defensive armament in cockpit	FuG 220
A-6	A-2 without armor	DB 603 E	2 x MG 151/20 in ventral tray 2 x MG 151/20 in wing roots	FuG 220
A-7/R1	prototype aircraft V25	DB 603 G	2 x MG 151 and 2 X MK 103 in ventral tray 2 x MK 108 in wing roots	FuG 220
A-7/R2	production version, prototype aircraft V26	DB 603 G	2 x MG 151/20 and 2 x MK 108 in ventral tray 2 x MK 108 in wing roots	FuG 220
A-7/R3	pre-production series for B-1 version, prototype aircraft V27	DB 603 G	2 x MG 151/20 in ventral tray 2 x MG 151/20 in wing roots 2 x MK 108 *schräge Musik*	FuG 220
A-7/R4	R2 with reduced armament	DB 603 G	2 x MG 151/20 in ventral tray 2 x MG 151/20 in wing roots	FuG 220
A-7/R5	Mosquito-hunter with methanol-water injection	Jumo 213 E	2 x MG 151/20 in ventral tray 2 x MG 151/20 in wing roots	FuG 220
A-7/R6	improved A-2, prototype aircraft V18	Jumo 222 A	4 x MK 108 in ventral tray 2 x MG 151/20 in wing roots	FuG 220

Radio Equipment Installed in the He 219 Series

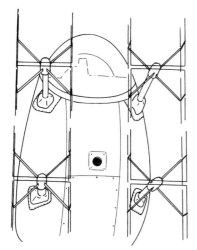

I. FuG 202 BC on He 219 prototypes (V) and pre-production (0) aircraft.

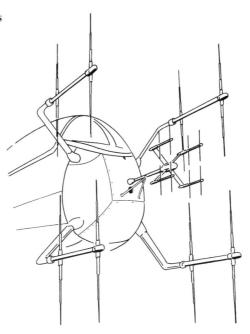

II. FuG 220 SN-2 with FuG 212 central antenna on He 219 pre-production aircraft (Zero-Series). Beginning roughly in the autumn of 1944 the central antenna was deleted.

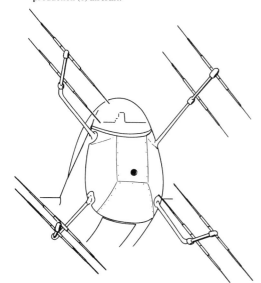

III. FuG 220 SN-2d with *Streuwelle VI* (dispersal waveband) on He 219 A-2 to A-7.

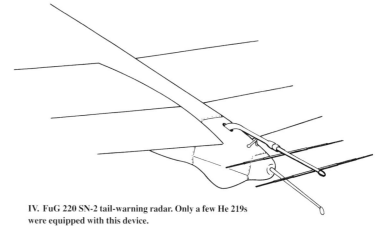

IV. FuG 220 SN-2 tail-warning radar. Only a few He 219s were equipped with this device.

V.   The radar operator's position in the He 219 A.

1 SG 212 or SG 220 R radar scope with visor
2   FBG 213 control device for FuG 220
3   FuG 10 P receiver E 10a K
4   FuG 16 ZE control box
5   Schk 17a switchbox
6   FuG 10P transmitter S 10 K
7   AFN2 beacon cross-pointer device
8   FBG 2 control device for E BL 3 F instrument approach receiver
9   SG 220 radar scope with visor
10   Schk 213 switchbox for FuG 220
11   FuG 10 P beacon receiver EZ 6 (lang)
12   FBG 3 control device for FuG 10 P
13   FuG 10 P transmitter S 10 L
14   PTK/p 2 D/F repeater compass

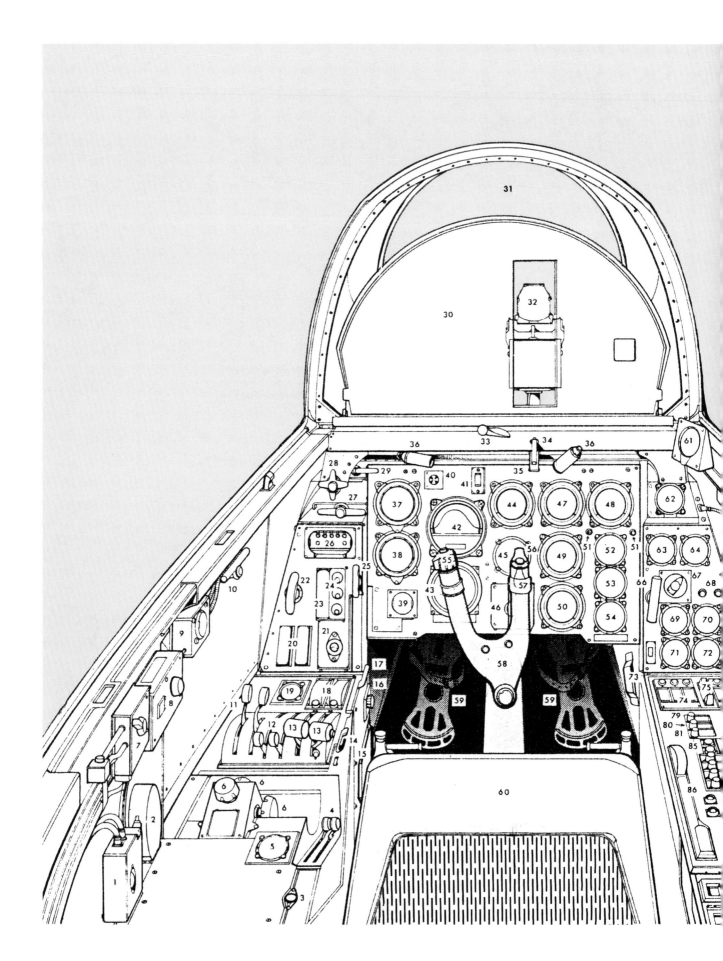

He 219A Pilot's Cockpit

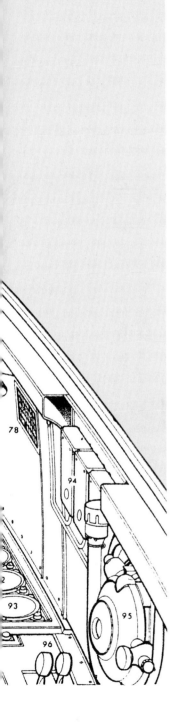

1   Heated flying suit plug-in
2   Enclosure for canopy emergency jettison safety fuse
3   Compressed air tank filler
4   Cooling gills emergency adjustment levers
5   Pressure-oil double pressure gauge (port and starboard)
6   Trim controls
7   FuG 17 (voice radio) control box
8   Intercom junction box
9   Revi gunsight dimmer switch
10   Cockpit ventilation control lever
11   RPM fine correction lever
12   Fuel tank selector switches
13   Throttle levers
14   Undercarriage brake locking lever
15   Map case
16   Cockpit heating regulator
17   Rudder ground lock handle
18   Magneto switches
19   Landing flap position indicators
20   Propeller pitch switches
21   Start-assist cable plug
22   Emergency flap-lowering lever
23   Undercarriage switches
24   Landing flap switches
25   Lever for lowering armored visor
26   Indicator light panel (12)
27   Main undercarriage emergency lowering lever
28   Nosewheel emergency lowering lever
29   Inner (armored) windscreen washer
30   Armored visor (early models)
31   Inner (armored) windscreen (later models)
32   Revi 16
33   Windscreen wiper control lever
34   Revi anti-glare screen switch
35   Revi night filter screen switch
36   Instrument panel lighting
37   Airspeed indicator with Fl 22241 altitude compensator
38   Fl 22322 fine/coarse altimeter
39   Emergency turn and bank indicator / later blanking cover
40   Visual indicator for Fl 32530-7 airspeed indicator
41   Autopilot emergency switch
42   Fl 22411 turn and bank indicator / artificial horizon
43   Fl 23371 repeater compass
44   Fl 22384 variometer
45   Fl 27002 indicator
46   Windscreen demist/clean manual control
47   Fl 20269 RPM gauge (port)
48   Fl 20269 RPM gauge (starboard)
49   Fl 20556 double manifold pressure gauge
50   FuG 101 (radio altimeter) gauge (later replaced with blanking cover)
51   Propeller feathering indicator lamps
52   Propeller pitch double indicator
53   Fl 20572 fuel pressure gauge (starboard engine)
54   Fl 20572 fuel pressure gauge (port engine)
55   Transmit button
56   Fuselage gun-firing button (right control horn)
57   Wing gun-firing trigger (rear face of horn)
58   Control column
59   Rudder pedals and brakes
60   Pilot's ejection seat
61   Course setting instrument (later replaced by clock)
62   Fl 23885 clock / later replaced with blanking cover
63   Fl 20723 oxygen supply meter
64   Oxygen pressure gauge
65   FK 38 Fl 23233 standby compass
66   Canopy jettison handle
67   Fuel selector switch
68   Temperature warning lamps
69   Fl 20341 coolant temperature gauge port engine
70   Fl 20341 coolant temperature gauge starboard engine
71   Fl 20341 engine oil temperature gauge port engine
72   Fl 20341 engine oil temperature gauge starboard engine
73   Compass deviation card holder
74   SZKK6 ammunition counters
75   Weapons safety switches
76   Phosphorescent lights
77   Starboard console lights
78   Ejection seat instruction plate
79   Landing light switch
80   Navigation light switch
81   Phosphorescent light switch
82   Gyro control monitoring switch

He 219 *Werknummer Summary*

219000	first four prototypes
190000	prototypes and pre-production series
190100	prototypes and pre-production series
190200	prototypes and pre-production series
290000	A-2 series
290100	A-2 series
290200	A-2 series
310000	A-7 series
310100	A-7 series
310300	A-2 series
420000	A-5 series

He 219 Production Numbers

Quarter	1942	1943	1944	1945
1st	4	26	62	
2nd	6	27	6	
3rd	14	38		
4th	2	26	63	
Total	2	50	154	68 = 274

Memories of a Night Fighter Pilot

16 January 1945, Münster/Handorf night fighter base. The missions were hopeless and depressing and most were also unsuccessful. Even before the order to cockpit readiness came, the "duty long-range fighter" was buzzing around the base. When you heard the sound of its Rolls-Royce engines, that typical loud buzz, then you knew that you would be driving out to the dispersals within half an hour at most.

In the bus that brought you to the machines there was deathly silence during the drive. Everyone struggled with his own thoughts in an effort to prepare for the coming mission, thoughts of the Mosquitoes, which all too often made you, the night fighter, the hunted, and for which you had so far failed to come up with any really effective remedy. The process of climbing into our Heinkel, the putting on of inflatable dinghy and parachute, helmet and throat microphone and the fastening of the harness was also done in oppressive silence. Then came the waiting with stage fright, which we also referred to as the "great blue funk".

The red flare signaled "cockpit readiness cancelled," then the inner tension went away. The bus came and collected the crews. The drive back to the command post was loud and boisterous, the pressure had given way to a forced cheerfulness. Steam had to be blown off: we had gotten away once more! If the green flare was fired it meant the order to take off. The activity which began involuntary forced the blue funk and the fear of takeoff. And when the machine had finally lifted off and the motors were droning in their pulsing rhythm, then all unrest was forgotten. In spite of the difficult circumstances night flying never failed to fascinate all over again.

I. Gruppe had shrunk to about a dozen crews, barely *Staffel* strength. It was rare that more than ten machines took off on a night mission, usually less, and of these half either returned immediately after take-off or were forced to land within the next half hour on account of malfunctions or problems. In the majority of cases it was the on-board electrics which failed. The aircraft sat in camouflaged blast pens at the edge of the airfield or in surrounding woods, unprotected against rain and storm, ice and frost. Condensation buildups led to an increasing number of shorts or bad contacts in the electrical system or to the malfunctioning of other important equipment, in particular the radio and DF systems, even though a maintenance test flight was carried out every two or three days to "fly the machine dry." In the fifth winter of the war the wiring was obviously no longer made from the best materials. It had happened that the guns suddenly fired when the undercarriage was retracted after take-off and could not be stopped by activating the appropriate circuit breaker. As well, in an attack on an enemy bomber the guns failed to fire when the button on the control column was pressed and the landing light came on instead. The pilot to whom this happened related, "Never before had I beat it so quickly, but the Tommy was obviously also so frightened that he forgot to press the button. What luck!"

Also depressing was the number of enemy night fighters. The Mosquito was an outstanding machine and the Fishpoint radar with which the night fighter version was equipped was unsurpassed, while the SN 2 which equipped the German night fighters at this stage of the war was constantly jammed. One of ours got it almost every week. Aircrew casualties were relatively low since the He 219's ejection seats made abandoning the aircraft a simple matter (assuming they had not been shot up). Pulling the familiar red lever blew the pilot and observer out of the aircraft (ninety and sixty atmospheres of pressure, respectively).

Their *Gruppe* had received the order to take off at about 1900 hours on that January night—it was very cold and the snow lay unusually deep. Eight machines took off from Münster-Handorf and were guided towards the enemy by Y-control.

"Eagle 98 from Wagon Wheel, fly heading one-nine-zero, Height four-five." The radio operator acknowledged: "Roger, heading one-nine-zero, out." "But not at that altitude," declared the pilot, "there are too many Mosquitoes lurking about there."

He made an instrument turn onto the assigned course and climbed the aircraft as quickly as possible to an Height of nine-thousand meters. In purely flying terms it was a very uncomfortable altitude. The reduction in air pressure made the control forces uncomfortably light; it was no longer flying, it was more like wallowing. The machine immediately lost height in every turn and the vertical speed indicator showed a descent rate of seven to eight meters per second. On the other hand one was relatively safe from the Mosquitoes. While on fighter missions they usually operated at heights between two- and three-thousand meters. Their Fishpoint radar could monitor a cone-shaped area above their machine, and the most dangerous heights were between three- and six-thousand meters. The great height offered another advantage: if one identified the "scene of the action" it was possible to turn height into speed by diving and reach the area of activity quickly. The attacking bombers usually operated at altitudes between four- and five-thousand meters.

The situation in the air was unclear. They were sent here and there, the whole time remaining over the east of the Ruhr Region. Radio communications with the control center were exceptionally bad. Jamming began as soon as the radio operator or the ground station had spoken ten words. As a result, radio traffic was limited to the briefest messages.

They must have been somewhere east or northeast of Cologne when the pilot's oxygen equipment failed. He noticed it immediately and without hesitating closed the throttles and put the machine into a steep dive. He knew very well that at that altitude he would be dead in two to two and a half minutes at most. The radio operator cried out, surprised by the sudden dive. "What's going on?" The pilot informed him of the oxygen system failure and that he would prefer to descend to an altitude of 1 000 to 1 500 meters in order to get beneath the Mosquitoes.

"Don't go too low. Most of the transmitters are jammed; the lower you go the more difficult it will be to get a good bearing. Stay at four-thousand until I know where we are and can give you a course for Münster. Then you can proceed as you wish." When the altimeter showed three-thousand he leveled off the machine, trimmed it out and took up a north heading. Home base had to lie somewhere in a northerly direction.

He knew very well that he was at an altitude which must be thick with Mosquitoes. Based on his own experiences in shooting down enemy aircraft he had come up with a special system for such emergencies: using the stopwatch, he flew on the desired heading for just three minutes. He knew that it usually took a night fighter three to four minutes to work its way into firing position or to reduce its speed to match that of its target before attacking.

After flying straight and level for three minutes he made a thirty- to forty-degree instrument turn to the right to simulate a normal course change. If he had been flying the pursuing night fighter he would have veered left to wait for the completion of this apparent course change and then commenced his attack. With the command "now!" he instructed his radio operator to scan the airspace behind their machine while he abruptly hauled the aircraft out of its shallow right turn into a steep left-hand turn. The radio operator should see any Mosquito sitting behind them in attack position. The unexpectedly quick course change would make it difficult for the Mosquito to get off a shot; even if its pilot, surprised by the sudden movement of his target, had pressed the buttons as a reflex action, the burst would surely have missed.

They continued north, repeating the defensive maneuver every three minutes. All the while the radio operator tried to tune in an unjammed transmitter in order to take a bearing. After several tries he succeeded in getting a cross-bearing from a powerful beacon in the East Eifel. He then tried to pick up Münster-Handorf's radio beacon. Just before the end of another three-minute period the pilot advised that he was about to commence the next maneuver.

"Wait a moment. I have just got Handorf, just a few more seconds." "Hurry up!" urged the pilot. Barely thirty seconds had passed when explosions shook the machine. The control column was ripped from his hands and with a snap the noise in the earphones of his helmet went dead. The intercom was dead.

"Long-range night fighter!" screamed the radio operator. The pilot could not hear above the roar of the engines, but he did not need to be told. Then he saw a Mosquito fly past his machine on the right, about twenty meters higher. "What an amateur," he thought, "probably never heard of equaliz-ing speed. Now I'll show you what it's like to be shot down." With his left hand he flipped the weapons switch. With his right he reached for the stick, his index finger felt for the firing button. As he tried to raise the nose of his machine to place the Mosquito in the crosshairs, he realized that the controls were limp. He could move the column forward and backward like a pendulum with no reaction from the machine. The elevator cable had apparently been severed in the attack. Luckily the ailerons were still intact.

The machine began to rear up. It climbed, lost speed and stalled. In the process it regained speed and the game began again. He tried to use the throttles to compensate for this up and down, but to little avail. Then he tried using the trim tabs as well, however they could not function as the control cables had been shot away. He realized that he would not be able to land the machine. They would have to bail out somewhere, but he wanted to delay this as long as possible. The longer he could keep the aircraft in the air, the closer they would be to their home airfield. The aircraft's movements were costing altitude. When the needle of the altimeter reached the 1 000 meter mark he knew that the time to bail out had come. He closed both throttles and shouted into the silence, "Get out – out!" The aircraft immediately stalled. He shoved the throttle levers forward again to level off the aircraft. The ejection seat safety lever was up, and he placed his feet in the "stirrups". Then he disconnected the throat microphone and took off his helmet. He pulled over his head the cord on which he had hung the flashlight around his neck and simply let the flashlight fall into the cockpit. After bailing out, when he was being tossed around by the slipstream, it might have caused him a black eye.

Having completed his preparations to bail out, he jettisoned the canopy and waited for the bang that indicated that his radio operator had ejected. When he heard this, he immediately banged on the ejection seat's firing button with his fist and was blown out of the machine. He wanted to come down as close to his radio operator as possible. Below it was dark and very cold. There was deep snow and it was impossible to say when and where one would land and whether it would be necessary to help the other. The slipstream seized him immediately and tossed him about. He released the harness and separated from the seat, then he waited a few seconds and pulled the parachute handle. When the parachute opened he immediately pulled his flare pistol from the right pocket of his Channel Pants and loaded it with a parachute flare from the ammunition belt he had placed around the shaft of his fur-lined boot. In the bright light produced by the magnesium capsule hanging beneath the handkerchief-size parachute, he saw his radio operator's parachute below to his right. The wind was carrying it slowly over a snow-covered wood toward a circular clearing surrounding a village.

He fired two more parachute flares, brightly illuminating the landing site. He drifted over the narrowest part of the clearing. On the other side, toward which he was drifting, he saw very tall fir trees, while on his side there were deciduous trees. It occurred to him that it would be easier to climb down from a deciduous tree than a fir. And so he pulled the parachute around a little, reached up with his arms and pulled down on the left parachute harness, to which were attached half of the thirty-two parachute lines. The parachute fell away to the left. He hoped to land in the clearing, but he had misjudged slightly, the parachute had slid rather too far to the left. He released the harness, hoping that the wind would carry him back toward the clearing. But then branches snapped under his weight and he was hanging in the middle of the crown of a mighty oak. The silk of his parachute covered the crown of the tree like a huge white flower. This was the third time he had come down by parachute and this was the softest landing ever.

He swung himself onto the nearest branch, and when he was on firm footing he pulled the quick release handle and released the parachute harness. The parachute flares had gone out by now and he fired another into the air. In its light he saw that his oak was standing on the steep bank of a small valley. Below there was a stream and beside it a narrow road which led over some sort of bridge directly into the village. His radio operator had come down on the road in the middle of the village. When he fired the flare he heard the latter shout, "Come down into the village—but be careful and don't fall into the brook!"

He climbed down the tree, branch by branch. His right knee pained fiercely and seemed to be swollen. Something must have happened but he did not know what and he also could not remember when this might have happened. He reached the bottom branch, and on looking around he realized that he was still quite high above the ground. As well, the thick trunk did not appear to be very well suited for climbing. He took the inflatable dinghy from his back, blew it up by unscrewing the small compressed air bottle, and let it slide down the trunk onto the snow. "It can't hurt to have a cushion in case of a fall," he thought. The idea of climbing down with his injured knee worried him. Finally he thought of his remote ancestors, who according to Darwin must have been accomplished climbers, and began to climb down. He clasped his arms and legs around the trunk—his knee causing him terrible pain—and slowly slid downwards. It was lucky that he still had his flying gloves. As he slid down it occurred to him that he had forgotten something: he should also have inflated his life vest and thrown that down too. The compressed air bottle for inflating it was hanging in front of his left hip. He felt its pressure as he slid down and hoped: "Hopefully it will remain peaceful." But it was too late to do anything about it. He continued sliding and then heard a hissing sound! Rubbing against the coarse bark had caused the valve on the compressed air bottle to open and inflate his life vest. Both bladders in front of his chest suddenly inflated, pushing his body away from the tree. His arms and legs could no longer grasp the trunk and he dropped like an elevator. The deep snow and his dinghy, on which he landed in the sitting position, softened the impact.

He stumbled down the bank, the deep snow over the tops of his fur-lined boots. His radio operator waited for him at the bridge. Just beyond the bridge was an inn. They knocked on the door and were let in.

It turned out that they had come down in a village called Laggenbeck.

The villagers had been very worried: they had heard aircraft sounds nearby, the crack of the flare pistol and finally the "Christmas Trees" of the pyrotechnics. Fearing an air attack, most villagers had fled into their cellars. Now they were happy to have escaped once again. And they had a pilot and radio operator in their midst.

After phoning their command post from the inn to relate their whereabouts, they were taken to a farmhouse where they were treated like princes. Early the next day a car picked them up and returned them to their *Gruppe*. The pilot's knee was so badly swollen that he had to rip open the seam of his uniform breeches. This did not prevent them from burning the midnight oil, however.

Test Report by RAE Pilot Eric Brown
(abbreviated version)

When I first encountered the He 219 at Grove, slightly less than two years after its auspicious debut at the hands of *Major* Streib, my primary concern was the Arado Ar 234 B and I could spare no time for more than a cursory examination of the Heinkel fighter on the ground. Five He 219s were eventually flown to the Royal Aircraft Establishment's captured enemy aircraft collecting point at Schleswig's airfield for ferrying to Farnborough, however, and I was soon to have an opportunity to fly three of these. The aircraft flown to the UK comprised four examples of the first production model, the He 219 A-2, and a single He 219 A-5, which was, in fact, the He 219 V11 rebuilt to A-5 standards after suffering extensive damage in an accident. Although we carried out no specific performance or handling tests on the He 219 at Farnborough, we did some tests on certain items of its equipment which were of particular interest and I made a number of ferry flights with the aircraft during the course of which there was plenty of opportunity to assess its flying qualities.

From the pilot's viewpoint, the most impressive feature of the He 219 was its cockpit, which, at a considerable height from the ground, was reached by means of a single-sided ladder which, after the crew members were aboard, was released by a member of the ground staff to pivot upward and rearward for stowage in a slot on the underside of the fuselage. The canopy was a large, one-piece affair which hinged to starboard and offered an excellent all-round view. The cockpit itself was roomy and comfortable, and the flight instruments were arranged in the classic 'T' formation, with the engine instruments grouped to the right of the main panel. The pilot and radar operator were seated back to back on compressed-air ejection seats and, as already

indicated, every item of equipment that a night fighter could have at this point in time was included. Indeed, that the *Reichsluftfahrtministerium* recognized the merits of the He 219 cockpit is indicated by the fact that, at one stage, serious consideration was apparently given to the possibility of grafting the entire nose on to the Junkers Ju 388.

The He 219 was always intended for the Daimler Benz DB 603 G engine rated at 1,900 h.p. for takeoff and 1,560 h.p. at 24,200 ft (7 375 m), but non-availability of this power plant when the first production airframes began to roll off the Vienna-Schwechat assembly line in the autumn of 1943 necessitated the substitution of the DB 603 A rated at 1,625 H.P for takeoff, 1,850 h.p. at 6,890 ft (2 100 m) and 1,625 h.p. at 18,700 ft (5 700 m), this model being designated the He 219 A-2. It was this version of the Heinkel fighter that I was to fly on several occasions, these flights including the ferrying of the He 219 A-2 (*Werknummer* 290126) from Farnborough to Brize Norton on 21 August 1945, delivering another A-2 (*Werknummer* 310109) to Abingdon nine days later and then to Brize Norton on the following day and, finally, on 19 October, collecting a third A-2 (*Werknummer* 310106) from Tangmere and ferrying it to Farnborough. I found no opportunity to fly the later DB 603 G-powered He 219 A-5 but imagine that it did not display any markedly different handling or performance characteristics to those of the A-2.

Starting the Db 603 A engines was simple. The fuel cocks were selected to tanks 2 and 3 (which were the center and rear of the three fuselage tanks) and the fuel pump for the requisite engine was turned ON. The throttle was opened about a quarter of its travel when a noticeable resistance could be felt, the magneto switches then being set to positions M1+2. Normally, the external electrically-operated inertia starters were used, although a similar internal system was available. The starter handle was depressed for 10-20 seconds, released and then pulled out for engagement, simultaneously being turned to the left to prime the engine if cold. Once the engine fired the revolutions were kept below 1,200 until oil and fuel pressures registered, when it could be warmed up at 1,500 rpm for three minutes before power checks were made. Ignition tests were made at 2,000 rpm.

Before commencing taxiing all trimmers were set to ZERO, the cooling gills were opened and the ejection seat air pressures were checked, that for the pilot's seat being 1,138 lb/sq in (80 kg/cm2) and that for the radar operator being 711 lb/sq in (50 kg/cm2). The Heinkel was very easy to handle during taxiing but the brakes—which were very effective—had to be used generously in crosswinds. Pre-takeoff checks included setting the airscrew pitch at 12.15 o'clock on the indicators and lowering the flaps to the START position. The takeoff run was a lengthy 1,700 yards (1 555 m) using full takeoff power of 2,700 rpm and 1.4 *atas* (20.6 lb) of boost.

I have read German reports that, fully loaded, the He 219 enjoyed an ample surplus of power and that an engine cutting immediately after takeoff or during the approach presented little danger. There was, it is said, an instance of a pilot making an emergency takeoff on one engine with his undercarriage locked in the "down" position and with flaps fully extended! If there is any truth in this last report, I can only say that for this extraordinary feat the aircraft must have been equipped with JATO and have had a very long runway indeed! In my view, the Heinkel fighter—certainly in its He 219 A-2 version—was decidedly *underpowered*. An engine failure on takeoff

must have been a *very* nasty emergency to handle at night as, below 137 mph (220 km/h) the aircraft was difficult to hold straight and, combined with the sink as the undercarriage came up, this meant that there was a critical area between 50 ft (15.20 m) and 300 ft (91.50 m) on climbout.

Unstick speed was 106 mph (170 km/h) and it was possible to commence raising the undercarriage at 50 ft (15.20 m) but not lower because of the previously-mentioned sink. As speed built up to 155 mph (250 km/h) the flaps could be raised at 500 ft (152 m), this being accompanied by a noticeable sink, and the aircraft could then be settled in a steady climb at 186 mph (300 km/h) with 2,500 rpm and 1.3 *atas* (19.1 lb) of boost. Once settled in the climb, the excellent stability characteristics of the He 219 became evident. The best rate of climb was obtained by letting the initial speed of 186 mph (300 km/h) decay slowly as altitude was gained until it dropped back to 174 mph (280 km/h) at 32,810 ft (10 000 m). The rate of climb was certainly unimpressive.

When the required cruise altitude was attained, the cooling gills were closed and the engines throttled back to 2,300 rpm and 1.2 *atas* (17.6 lb). If flying for endurance then power was further reduced to 2,000 rpm and 1.05 *atas* (15.4 lb). A full power run at 20,000 ft (6 096 m) revealed somewhat sluggish acceleration and a top speed of 378 mph (608 km/h), which was somewhat below the German handbook figures.

Operation of the fuel system required tanks 2 and 3 to be used initially until half-empty (i.e., about 220 Imp gal/1 000 liters remaining), thereafter switching to the forward or No 1 tank until its contents were exhausted. The cabin heating and de-icing systems were extremely effective and the autopilot was easy to operate and appeared reliable. The He 219 was, without a doubt, an excellent all-weather aeroplane.

Landing preparations were straightforward. The cowling gills were opened, the airscrew pitch set to 12 o'clock and, at 186 mph (300 km/h) the flaps were lowered to START. The undercarriage was lowered at 168 mph (270 km/h) and final turn-in made at 155 mph (250 km/h). Full flap and fuel pumps were selected on the final approach and speed held at 140 mph (225 km/h), reducing to 124 mph (200 km/h) near the airfield boundary. The touchdown at 99 mph (160 km/h) was very easy with the tricycle undercarriage, but the nosewheel could not be held off the ground for very long and so the landing was fast and strong braking was called for. It was vital that, before landing, a check was made of the brake pressure which should have been at least 853 lb/sq in (60 kg/cm2). If below that figure, a button alongside the gauge had to be pressed until the pressure built up and maintained itself. The landing run in zero wind was about 710 yards (650 m).

It was apparent in the landing condition that the lateral control afforded by the spring tab ailerons was sluggish in effect and hence turns with the flaps fully lowered were not to be recommended. In gusty conditions the Heinkel was unpleasant to handle laterally on the final approach.

From my experience with the He 219 A-2, I would say that the Heinkel fighter's reputation was somewhat overrated. It was, in my view, basically a good night fighter in concept but it suffered from what is perhaps the nastiest characteristic that any twin-engined aircraft can have—it was underpowered. This defect makes takeoff a critical maneuver in the event of an engine failure and a landing with one engine out can be equally critical.

Sources

1940;
Aktennotiz P1055, EHF, 2.10.1940,
Abschrift P1055, EHF, 3.10.40,
Einschreiben P1055, EHF, 8.10.40,
Aktenvermerk P1055, EHF, 24.10.40
Aktenvermerk P1055, EHF, 23.11.40
Aktenvermerk P1055, EHF, 28.10.40
Terminstand , EHF, 16.12.40

1941;
Fernschreiben Typenbezeichnung 227/1,
    EHF 11.1.41
Fernschreiben Attrappenbesichtigung 35/2,
EHF 3.2.41
Aktenvermerk Projekt He219, EHF, 13.2.41
Aktenvermerk He219, EHF, 17.2.41
Aktenvermerk He219 Projektumstellung,
    EHF, 25.2.41
rief an Dir.Lusser, EHF, 26.2.41
nvermerk Projekt He219, EHF, 14.3.41
Mitteilung, Kanzelattrappe, EHF, 17.3.41
Aktenvermerk Projekt He219, EHF, 27.3.41
Aktenvermerk 3 Sitzer He 219, EHF, 10.5.41
Aktenvermerk Periskop He219, EHF, 14.5.41
Aktenvermerk Besprechung im RLM, EHF, 3.6.41
Protokoll Besprechung über He 219,
    EHF, 23.6.41
Fernschreiben 634/6, EHF, 25.6.41
Fernschreiben 317, EHF, 1.7.41
Aktenvermerk Projekt He219, EHF, 11.7.41
Aktenvermerk Projektbesprechung He219,
EHF, 18.7.41
Infobrief an Gen.Obst Udet, 19.7.41
Aktenvermerk He219 Nachtjäger, EHF, 31.7.41
Aktenvermerk Ausrüstung, EHF, 1.8.41
Aktenvermerk Projektumstellung auf Außen-
    motoren He219, EHF, 14.8.41
Fernschreiben 396/8, EHF, 16.8.41
Fernschreiben 504/8, EHF, 20.8.41 und Antwort
Fernschreiben 308/47, EHF, 25.8.41
Aktenvermerk Nr.68 Projektbesprechung He219,
    EHF, 2.9.41
Protokoll Besprechung mit Generalstab,
    EHF, 13.10.41
Mitteilung an Ing. Schwärzler/Günter,
    Dr.Heinkel, 25.10.41
Reisebericht 2646, EHF, 1.11.41
Aktenvermerk Projektbesprechung He219,
EHF, 5.12.41

1942;
Fernschreiben He 219, EHF, 27.1.42
Fernschreiben He 219, EHF, 11.2.42
Reisebericht 2893, EHF, 20.2.42
Terminfestlegung an Daimler-Benz AG,
        EHF, 11.3.42
Fernschreiben-Antwort an Herrn Beu, Dr.Heinkel,
26.3.42
Brief an Herrn Mescchkat, Dr.Heinkel, 26.3.42
Mitteilung 22/42, EHF, 27.3.42
Antwortschreiben Motorenlieferung,
    Daimler Benz AG, 27.3.42
Fernschreiben 735, EHF, 31.3.42
Mitteilung an Konstruktionsbüro,
Dr.Heinkel, 2.4.41

Fernschreiben Antwort, EHF, 7.4.42
Mitteilung 54/42 Motorenbeschaffung f. He 219,
    EHF, 15.5.42
Mitteilung 314/42 zur Besprechung b. DB,
EHF, 27.5.42
Mitteilung 133/42, EHF-Entwurf, 1.6.42
Mitteilung 348/42, Großserienplanung He177
    und 219, EHF, 26.6.42
Aktennotiz Dr.Heinkel zur Werksverlagerung,
    EHF, 7.7.42
Mitteilung 354/42 Motorenlieferung, EHF, 8.7.42
Brief an das RLM, EHF, 30.7.42
Fernschreiben 275 Baukosten He 219,
    EHF, 1.8.42
Fernschreiben 95/8 Dringlichkeitseinstufung,
    EHF Berlin, 4.8.42
Besprechungsniederschrift im RLM,
    EHF, 22.8.42
Mitteilung zum Leitwerk He 219,
Dr.Heinkel, 9.9.42
Anforderung für Fahrwerke He 219, EHF, 9.9.42
Aktenvermerk He 219 Lieferplan, EHF, 14.9.42
Brief an Prof. Dr.Heinkel betr. Aufgabenreihen-
    folge, RLM Generalinsp.d. Luftwaffe Milch,
    15.9.42
Mitteilung zur Erstflugvorbereitung,
    EHF, 24.9.42
Mitteilung 253/42 Zentralleitwerk,
    EHF, 17.10.42
Berichtigung zum Lieferprogramm 222,
    EHF, 2.11.42
Aktenvermerk zue Einstellung Zentralleitwerk,
    EHF, 5.11.42
Fernschreiben Erstflug He 219,
EHF, 6.11.42
Mitteilung 65/42 Beanstandungen zum Erstflug,
    EHF, 6.11.42
Infobrief an GFM Milch zum Erstflug He 219,
    EHF, 7.11.42
Fernschreiben Unfall V1, EHF, 9.11.42
Mitteilung He Projekte, EHF, 15.11.42
Geheime Aktennotiz zur Besprechung mit
    GFM Milch, EHF, 16.11.42
Mitteilung 52/42 zur Flugbeurteilung He 219,
    EHF, 17.11.42
Fernschreiben 71566 an Hirthmotor Stuttgart,
    EHF, 18.11.42
Fernschreiben 71573 an Hirthmotor Stuttgart,
    EHF, 18.11.42
Fernschreiben 71745 an Prof. Dr.Heinkel,
EHF, 19.11.42
Fernschreiben 71907 an Prof. Dr.Heinkel,
EHF, 20.11.42
Fernschreiben an EHF, Prof. Dr.Heinkel,
    23.11.42
Aktenvermerk 145/42 Stand der Versuchsflüge,
    EHF, 26.11.42
Fernschreiben 73509 Industrieprogramm 222,
    EHF, 26.11.42
Aktenvermerk 147/42 Flugerprobung, EHF, 2.12.42
Brief an GFM Milch, Prof. Dr.Heinkel, 20.12.42

1943;
Brief an Techn.Amt zur Arbeiterlage, EHF, 8.1.43
Aktenvermerk 155 Besprechung im EHF,

EHF, 13.1.43
Aktenvermerk 157/43 Stand der Versuchsflüge,
  EHF, 15.1.43
Aktenvermerk Besuch d. Maj.Streib,
  EHF, 16.1.43
Aktenvermerk 162/43 Anruf Gen.Kammhuber,
  EHF, 28.1.43
Brief an Gen.Kammhuber betr. He 219,
  Prof.Dr.Heinkel, 2.2.43
Brief an Prof. Dr.Heinkel, Gen.Kammhuber,
  5.2.43
Fernschreiben betr. Kanzelhaube, EHF, 16.2.43
Aktenvermerk Versuchsträger He 219,
  EHF, 18.2.43
Brief an Meister Richmann EHF betr. Prämie,
  Prof. Dr.Heinkel, 19.2.43
Fernschreiben 1038/2, EHF, 24.2.43
Fernschreiben 1057/2 Waffenlieferung und
  Triebwerksmangel, EHF, 25.2.43
Mitteilung 204/3 Probleme mit He 219 Fertigung,
  EHF, 3.3.43
Geheime Aktennotiz zur He 219 0-Serie,
  EHF, 6.3.43
Fernschreiben 11252 Fertigungsplanungen,
  EHF, 9.3.43
Fernschreiben 341/3 Erstflug für Kammhuber,
  EHF, 9.3.43
Fernschreiben 12744 He 277/He219,
  EHF, 9.3.43
Bericht zum Stand der Versuchsflüge,
  EHF, 10.3.43
Fernschreiben z. Waffeneinbau He 219,
  EHF, 14.3.43
Fernschreiben 7623 z. MK 108, EHF, 14.3.43
Aktenvermerk 38/43 Stand der Versuchsflüge,
  EHF, 14.3.43
Fernschreiben 7625 an EHF Nachtflug mit V3,
  EHF, 14.3.43
Aktenvermerk Personalmangel, EHF, 17.3.43
Kurzbericht 1100/43 z. Start- und Landemessung
  He 219, EHF, 18.3.43
Aktenvermerk 169/43 Vergleichsfliegen He219/
  Ju188, EHF, 23.3.43
Aktenvermerk He274/219, EHF, 24.3.43
Mitteilung 367/43 z. He 219, EHF, 27.3.43
Fernschreiben 592 z. Vergleichsfliegen,
  EHF, 25.3.43
Aktenvermerk 171/43 z. Vergleichsfliegen,
  EHF, 28.3.43
Bericht 15/43 z. Fahrwerk, EHF, 8.4.43
  und 26.3.43
Fernschreiben 354/4 z. Lieferlage DB 603
  Motoren, EHF, 9.4.43
Fernschreiben 13400z. Flugunfall V3,
  EHF, 19.4.43
Fernschreiben18475 z. Flugunfall V3,
  EHF, 20.4.43
Kurzbericht 1098/43 z. Start- und Lande-
  messung, EHF, 30.4.43
Flugeigenschaftsbericht 23/43, EHF, 30.4.43
Fernschreiben 916 z. Auslieferung der Versuchs-
  träger, EHF, 3.5.43
Fernschreiben 988 z. Beschuß V7, EHF, 3.5.43
Fernschreiben 1276 Vorfliegen v.d. Reichs-
  marschall, EHF, 6.5.43
Fernschreiben 1278 z. FuG 16 ZE,
  EHF, 6.5.43
Fernschreiben 1375 z. Stand der Versuchsflüge,
  EHF, 8.5.43
Fernschreiben 238/5 z. Holzflügel, EHF, 10.5.43
Geheime Aktennotiz z. Fertigungsprüfung
  in Wien, EHF, 13.5.43

Bericht He 219 V7 – V9, Techn.Ing. b.
  NJG1, 18.5.43
Bericht 25/43 z. Spannweitenvergrößerung,
  EHF 19.5.43
Bericht 110/43 z. Enteiosungsanlage,
  EHF, 25.5.43
Mitteilung 152/43 z. He 219 Geschwindigkeit,
  EHF, 26.5.43
Bericht He 219 V7 – V9, Techn.Ing. b.
  NJG1, 29.5.43
Mitteilung z. Vorfliegen v. Führungsstab,
  EHF, 30.5.43
Geheime Aktennotiz z. Vorfliegen v.
  Führungsstab in Venlo, EHF, 31.5.43
Mitteilung 156/43 z. Verlagerung 0-Serienbau,
  EHF, 2.6.43
Mitteilung 165/43 z. Liefertermin
He 219 V11/12, EHF, 7.6.1943
Besprechungsniederschrift z. Gesamt-
  entwicklung, EHF 10.3.43
Fernschreiben 167/43 z. Ersteinsatz He 219,
  XII. Fliegerkorps, 12.6.43
Fernschreiben 1129 z. Änderungsanweisung
V10, EHF, 12.6.43
Übersicht z. Terminlichen Ablauf Entwurf
  bis z. Serie, EHF, 22.6.43
Fernschreiben z. Anlauf Großserie He 219,
EHF, 29.6.43
Berichte z. Reichweitenvergrößerung mittels
Zusatzbehälter, EHF, 30.6.43
Aktennotiz z. Reichwweitenvergrößerung,
  EHF, 5.7.43
Fernschreiben z. Unfallbericht V2, EHF, 13.7.43
Protokoll z. Besprechung ü. Bremsschirm
  He 219, EHF 24.7.43
Protokoll z. Besprechung ü. Großserienfertigung
  He 219, EHF 27.7.43
Fernschreiben 58769 z. Einsatz V10,
  EHF, 27.7.43
Fernschreiben 7425 z. Einbau FuG 220,
  EHF, 10.8.43
Aktenvermerk TD47 z. Hohentwiel Einbau,
EHF, 16.8.43
Protokoll z. Staatssekretärbesprechung,
  EHF 20.8.43
Aktennotiz z. Übernahmetermin He 219,
  EHF, 24.8.43
Fernschreiben 8920 z. Abgabe He 219 A-01,
EHF, 27.8.43
Protokoll z. Besprechung ü. allgemeine
        Entwicklungsfragen, EHF 28.7.43
Fernschreiben 1254 z. Ausbringungsplan
  He 219, EHF, 31.8.43
Mitteilung ü. V10/V12 in Venlo, EHF, 31.8.43
Fernschreiben z. Stand He 219 m. FuG 220,
EHF, 2.9.43
Fernschreiben 1642 z. Stimmungen f.
  He 219, EHF, 6.9.43
Aktennotiz z. Übernahmetermin He 219,
  EHF, 24.8.43
Fernschreiben 8920 z. Abgabe He 219 A-01,
EHF, 27.8.43
Protokoll z. Besprechung ü. allgemeine
        Entwicklungsfragen, EHF 7.9.43
Fernschreiben 1254 z. Fertigungsablauf,
  EHF, 8.9.43
Brief an d. RLM z. Fertigungsablauf,
  EHF, 9.9.43
Techn. Bericht A-07, EHF, 10.9.43
Flugprogramm A-010 mit TL-Triebwerk,
  EHF, 22.9.43
Mitteilung z. Prämie f. A-010, Prof. Dr.Heinkel,
  23.9.43

Fernschreiben z. A-010, EHF, 24.9.43
3 Fernschreiben z. He 219 Maßnahmen,
  EHF, 29.9.43 -1.10.43
Reisebericht; Unfall A-03 bei Celle, EHF, 2.10.43
Aktenvermerk 503 z. schlechte Werkstattarbeit,
  EHF, 7.10.43
Reisebericht v. Front Erprobungsstelle Venlo,
  EHF, 6.10.43
Aktenvermerk z. Besuch b. NJG1, EHF, 14.10.43
Aktenvermerk 506 z. A-010 Arbeitsanweisungen,
  EHF 15.10.43
Reisebericht, Unfall A-04 bei Stendal,
  EHF, 23.10.43
Befund- und Prüfbericht 98/43, E-Stelle Venlo,
  25.10.43
Reisebericht, z. Besuch in Venlo, EHF, 25.10.43
Bericht He 219 A-010 TL, EHF, 25.10.43
Reisebericht, z. Besuch in Venlo, EHF, 26.10.43
Fernschreiben z. Übergabetermine,
  EHF, 27.10.43
Mitteilung 179/43 Lagebericht He 219 Fertigung,
  EHF, 28.10.43
Mitteilung 197/43 z. Fertigung Kabinendach,
  EHF, 31.10.43
Aktenvermerk 191/43 z. Beanstandungen b.
He 219, EHF 5.11.43
Fernschreiben z. Unfall A-010 TL, EHF, 13.11.43
Techn. Bericht 1133/43 z. Flugerprobung
  A-010 TL, EHF 19.11.43
Programm z. Überprüfungsflüge, EHF, 6.12.43
Fernschreiben z. Frontklar-Meldunen He 219,
  EHF, 8.12.43
Mitteilung z. Ablieferung einsatzklarer He 219,
  EHF, 9.12.43
Mitteilung z. Namensgebung, EHF, 14.12.43
Besprechungsniederschrift z. E-Programm
He 219, E-Stelle Rechlin, 30.12.43
Terminnennung zur Flugklarmeldung f.
  He 219 m. Jumo222, EHF, 30.12.43

1944;
Bericht 46 über Entwicklungsbesprechung m.
  RLM, EHF, 13.1.44
Aufstellungsübersicht der Versuchsträger,
EHF, 9.2.44
Entwurf z. Denkschrift Mosquito-Nachtjagd,
EHF, Febr.44
Brief an Dir.Frydag, Prof. Dr. Heinkel, 3.4.44
Protokoll z. he 219 m. 3 Mann Kanzel,
  EHF, 11.4.44
Anweisung Umbau V19, EHAG, 13.4.44
Stimmen der Front z. He 219, EHAG, 14.4.44
Mitteilung 554/44 z. Nachrüstung in Venlo,
EHF/Wien, 15.4.44
Aktenvermerk 3 Mann Kanzel, EHAG, 15.4.44
Mitteilung z. Verzicht auf Bau v. V23,
  EHAG, 17.4.44
Mitteilung z. Entwicklungsverlauf He 219,
  EHAG, 21.4.44
Mitteilung z. Anforderung v. 2 Umbaumaschinen
  f. 3-Mann-Kanzel, EHF/Wien, 8.5.44
Aktenvermerk z. Behelfsbomber He 219,
  EHAG, 15.5.44

Mitteilung z. Versuchsträger 3-Mann-Kanzel,
  EHAG, 17.5.44
Aktenvermerk z. Probeflug m. V34, NJGr.10
Werneuchen, 7.6.44
Reisebericht z. Beanstandungen der Truppe
in Venlo, EHAG, 7.6.44
Denkschrift He 219, EHAG, 13.6.44
Denkschrift He 219, Pr.Dr. Heinkel, 15.6.44
Aktenvermerk TD/27/44 He 219, EHAG, 20.6.44
Anlage z. Reisebericht 779/44 ü. Abschußerfolge
  NJG1, EHAG, 22.6.44
Fernschreiben z. 3-Mann Hc 219, EHAG, 29.6.44
Protokoll z. IIc 219 Weiterentwicklungen,
  EHAG, 13.7.44
Auftrag zur Baureihenumstellung A2 auf A5,
RLM, 15.7.44
Mitteilung 1/44 ü. Abschußerfolge NJG1 in Venlo,
  EHAG, 25.7.44
Aktennotiz z. Besprechung m. RLM ü. Leistungs-
  steigerung He 219, EHAG, 28.7.44
Aktenvermerk ü. 3-Mann Kanzel als Baureihe d.
  A5, EHAG, 31.7.44
Mitteilung z. He 219 m. Jumo 213/222,
  EHAG, 15.8.44
Abschrift d. Monatsmeldung ü. Serienbau,
EHAG, 31.8.44
Schreiben z. Hü211 – Verfügungstellung v.
Bauteilen He219, RM/Rüstung, 29.9.44
Industrielieferplan He 219, Sd. Ausschuß F3,
11.12.44

Bibliography

Geschichte der deutschen Nachtjagd, Gebhard Aders;
Motorbuch Verlag 1977
Luftfahrt International; Pawlas Archiv 1976
Jägerblatt:Gemeinschaft der Jagdflieger
He 219 Profile: Flugzeug Publikation GmbH, 1991
He 177, Griehl und Dressel; Motorbuch Verlag 1989
L.Dv.T.2219 A-0/FL.; RLM 1943/44
D. (Luft.)T.2219 A-0 Teil 2; EHAG 1944
D. (Luft.)T.2219 A-0 Teil 9D; EHAG 1944
Ersatzteilliste He 219 A-7; EHAG 1944
Modell Fan 1979
Stürmisches Leben, Ernst Heinkel; Mundus
Verlag 1954, AVIATIC 5, 1991
Profile 219, Richard Bateson; Profile Publications Ltd
1971
Warplanes of the Third Reich, William Green; Mac-
Donald Ltd. 1970
Flugzeug Fahrwerke, Günther Sengfelder; Motor-
buch Verlag 1979
Flugzeugbewaffnung, Schliephake; Motorbuch Ver-
lag 1977
Die nicht zurückkehrten, Werner Girbig; Motorbuch
Verlag 1970

Internet
http://www.nasm.edu/nasm/AERO/AIRCRAFT/
heinkel@219.htm
http://www.nasm.edu/AERO/AIRCRAFT/gtok.htm